SpringerBriefs in Cancer Research

For further volumes:
http://www.springer.com/series/10786

Hiroyuki Inuzuka • Wenyi Wei

SCF and APC E3 Ubiquitin Ligases in Tumorigenesis

Springer

Hiroyuki Inuzuka
Department of Pathology
Beth Israel Deaconess Medical Center
Harvard Medical School
Boston, MA, USA

Wenyi Wei
Department of Pathology
Beth Israel Deaconess Medical Center
Harvard Medical School
Boston, MA, USA

ISBN 978-3-319-05025-6   ISBN 978-3-319-05026-3 (eBook)
DOI 10.1007/978-3-319-05026-3
Springer Cham Heidelberg New York Dordrecht London

Library of Congress Control Number: 2014933521

Printed on acid-free paper

Springer is part of Springer Science+Business Media (www.springer.com)

Contents

Chapter 1
Introduction

Pengda Liu, Hiroyuki Inuzuka, and Wenyi Wei

Abstract Cancer, medically termed as malignant neoplasm, is characterized by uncontrolled cell division and growth, resulting from genome instability or dysregulated responses to physiological cues that regulate normal cellular processes including proliferation, survival, and differentiation. Intensive investigations over the past decades have demonstrated that cancer can be considered largely a genetic disease, and typically multiple genetic alteration events are required in order for primary cells to become malignant, including loss of tumor suppressor functions combined with gain of oncoprotein functions (Fukasawa, Nat Rev Cancer 7(12):911–924, 2007; Crusio et al., Oncogene 29(35):4865–4873, 2010).

Recently, there has been a wealth of literature demonstrating that F-box proteins, complexed with other essential components (Skp1 and Cullin) to form SCF-type of E3 ubiquitin ligase complexes, play a pivotal role in the development and progression of human malignancies (Nakayama and Nakayama, Nat Rev Cancer 6(5): 369–381, 2006; Welcker and Clurman, Nat Rev Cancer 8(2):83–93, 2008). Mechanistically, mounting evidence supports the notion that F-box proteins are involved in governing multiple cellular processes including cell proliferation, apoptosis, invasion, angiogenesis, and metastasis (Cardozo and Pagano, Nat Rev Mol Cell Biol 5(9):739–751, 2004). With many excellent studies in recent years regarding how F-box proteins contribute to human diseases such as cancer (Nakayama and Nakayama, Nat Rev Cancer 6(5):369–381, 2006; Welcker and Clurman, Nat Rev Cancer 8(2):83–93, 2008), now is a pertinent time to review our current understanding of how F-box proteins, including the well-established Fbw7, Skp2, and β-TRCP, are involved in tumorigenesis by controlling cell growth and apoptosis, regulation of invasion and metastasis, display of stem cell features, and establishment of drug resistance. Moreover, we also review the underlying mechanisms by which

P. Liu • H. Inuzuka (✉) • W. Wei (✉)
Department of Pathology, Beth Israel Deaconess Medical Center,
Harvard Medical School, Boston, MA 02215, USA
e-mail: hinuzuka@bidmc.harvard.edu; wwei2@bidmc.harvard.edu

H. Inuzuka and W. Wei, *SCF and APC E3 Ubiquitin Ligases in Tumorigenesis*,
SpringerBriefs in Cancer Research, DOI 10.1007/978-3-319-05026-3_1,
© The Author(s) 2014

1

F-box proteins are regulated, and how these pathways when disrupted can promote tumorigenesis.

In addition to SCF ubiquitin ligases, the Anaphase Promoting Complex/Cyclosome (APC/C, also called APC) is also a major ubiquitin ligase, which is a driving force in governing proper cell cycle progression, especially regulating timely transitions during mitosis, and entry into S phase (Peters, Nat Rev Mol Cell Biol 7(9):644–656, 2006). The APC consists of a core holoenzyme and an adaptor protein, either Cdh1 or Cdc20. Recent genetic and biochemical studies revealed that APC^{Cdc20} as a putative oncoprotein (Li et al., Mol Cell Biol 27(9):3481–3488, 2007; Manchado et al., Cancer Cell 18(6):641–654, 2010; Yin et al. Cell Cycle 6(23):2990–2992, 2007) while APC^{Cdh1} likely functions as a tumor suppressor (Garcia-Higuera et al. Nat Cell Biol 10(7):802–811, 2008; Li et al. Nat Cell Biol 10(9):1083–1089, 2008), yet the underlying molecular mechanisms by which these two lases exert their effects on tumorigenesis remain largely undefined. Moreover, studies from various groups have revealed an intensive crosstalk between the APC and SCF E3 ligase complexes in coordinating the timely cell cycle transitions. Hence, it is critical to summarize recent advances in our genetic and biochemical understanding of how various APC and SCF complexes and their regulators function in tumorigenesis, which will be useful in guiding the development of specific inhibitors targeting ubiquitin ligase function as novel anticancer treatments.

Keywords Ubiquitination • UPS • Proteasome • F-box • SCF • APC • Cullin • Phosphorylation • Cell cycle • Degradation • Tumor suppressor • Oncoprotein • Mouse model • Human cancer

1.1 Mammalian Cell Cycle Control

Aberrant cell cycle control is a hallmark of cancer making it critical to understand how the cell cycle progression is tightly regulated under normal conditions and how cancer cells evade this delicate regulation to drive uncontrolled cell growth. The mammalian cell cycle is driven by the regulated oscillating activities of cyclin-dependent kinases (Cdks) [11, 12]. Cdks are activated by association with their Cyclin partners that, as their name suggest, are expressed periodically during particular cell cycle phases. There are two groups of Cyclins, the G1-Cyclins including Cyclin E, Cyclin D, and Cyclin A [13] and the mitotic Cyclins that include Cyclin A and Cyclin B [14, 15]. Specifically, expression of both Cyclin A and Cyclin E peaks during the G1 to S phase transition and both of them complex with Cdk2 to promote S phase entry and DNA replication. On the other hand, Cyclin A or Cyclin B expression peaks during the G2 to M phase transition, associating with Cdk1 (also known as Cdc2) to govern mitotic transitions [16, 17].

In addition to activation through association with Cyclins, Cdks can be negatively regulated by Cdk inhibitors (CdkIs). CdkIs govern cell cycle arrest in response to various antiproliferative or stress signals such as growth factor deprivation, low energy status, DNA damage, or contact inhibition. Generally, two families of CdkIs have been characterized in mammalian cells based on their sequence homology, namely, the CIP/KIP family composed of p21, p27, and p57; and the INK family composed of p15, p16, p18, and p19. Induced expression of CdkIs in response to various stimuli leads to inactivation of Cdk kinase activities, resulting in cell cycle arrest either in the G1 or G2 phase. Progression through the cell cycle is tightly controlled by regulation of each transition through a variety of mechanisms, including expression of regulatory factors (i.e., Cyclins and CdkIs), posttranslational modifications (i.e., phosphorylation, acetylation, or methylation), and degradation of various regulatory proteins. In contrary to posttranslational modifications such as phosphorylation, which are reversible [18], destruction of cell cycle regulators by the ubiquitin-proteasome system (UPS) is irreversible, which assures the strict unidirectional operation of the cell cycle in a spatial and temporal manner.

1.2 Role of the UPS System in Cell Cycle Control

Ubiquitination and degradation of proteins by UPS governs diverse cellular processes such as cell proliferation, cell cycle progression, transcription, and apoptosis [19–23]. Protein degradation by UPS consists of two discrete processes (Fig. 1.1). The first process is the covalent attachment of the ubiquitin molecules to the targeted substrate protein, and the second process is the selective degradation of the ubiquitinated protein by the 26S-proteasome complex. UPS attaches ubiquitin molecules to substrate targets through a three-step enzymatic reaction cascade catalyzed by the ubiquitin-activating E1 enzyme, the ubiquitin-conjugating E2 enzyme, and the E3 ubiquitin ligase (Fig. 1.1). The E1 enzyme utilizes ATP to activate an ubiquitin molecule, which is subsequently transferred to the E2 enzyme, leading to the interaction between E2 and E3 enzymes. Then E3 binds to, and facilitates the transfer of the ubiquitin moiety to target proteins, marking them for degradation by the 26S proteasome in an ATP-dependent manner [2, 24, 25].

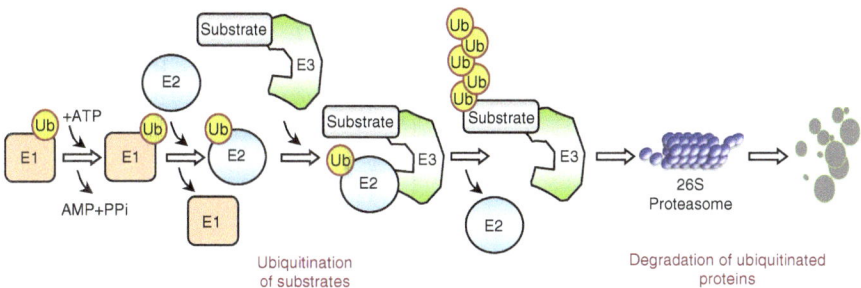

Fig. 1.1 A schematic illustration of the E1–E2–E3 cascade-mediated ubiquitin transfer process to a given substrate

Targeted proteins can be modified by monoubiquitination or polyubiquitination. Whereas monoubiquitination is rarely involved in degradation, polyubiquitination is involved in various cellular processes including degradation of the target protein through the 26S proteasome, which is dictated by how ubiquitin moieties are added to target proteins. The UPS assembles various polyubiquitin chains by conjugating the C-terminal glycine residue of ubiquitin to any of the seven-lysine residues (K6-, K11-, K27-, K29-, K33-, K48-, or K63-linkage) or the amino terminus (M1 linkage) of the previously loaded ubiquitin molecule on the substrate(s) [19, 26]. Among all the available ubiquitin linkages, the cellular roles of K48- and K11-linked polyubiquitination are well characterized as degradation signatures for targeting substrates for 26S proteasome-mediated degradation [4, 20, 27]. In contrast, monoubiquitination, K63-linked or M1-linked polyubiquitination have been demonstrated to play important roles in regulating many cellular processes including endocytosis, cellular trafficking, protein localization, and DNA damage responses [19, 28–30]. Although utilization of other atypical polyubiquitin linkages such as K6, K27, K29, or K33 has been reported, their functions remain poorly understood [19] and warrant further investigations. In addition, because ubiquitination reaction is a reversible process by deubiquitinating enzymes (DUBs) [31–33], thus the final ubiquitination status of a given substrate is an outcome of a balance between E3 ligases and DUB activity. Ubiquitin-proteasome-mediated proteolysis regulates cell cycle progression via timely destruction of many cellular regulatory proteins including Cyclins and CdkIs, and thus is a critical mechanism governing faithful replication and segregation of cellular genetic material [12, 18, 34].

1.3   SCF and APC E3 Ubiquitin Ligase Complexes

As E3 ubiquitin ligases determine the substrate specificity for ubiquitination, there is no surprise that to date more than 600 E3 ligases have been identified in the human genome [35]. Based on protein sequence homology, the E3 ligases are divided into three major classes: the HECT (Homologous to the E6-AP Carboxyl Terminus) family, the PHD (Plant Homeodomain)/U-box family, and the RING (Really Interesting New Gene) finger family [36–38]. The HECT E3 ligases form a transient and covalent interaction with ubiquitin to mediate the transfer of the ubiquitin molecule to its substrates, while the RING finger and PHD/U-box domains only facilitate the transfer of ubiquitin molecules from the E2 enzyme directly to their substrates [36, 39–41]. The RING finger family is further subgrouped by single-subunit RING protein containing both the RING and substrate binding domains or multi-subunit RING complex. The Cullin-Ring Ligases (CRL-type of E3s) are the most well-characterized RING E3 ubiquitin ligases, and contains eight family members, namely, CRL-1, CRL-2, CRL-3, CRL-4A, CRL-4B, CRL-5, CRL-7, and CRL-9 [42, 43]. CRL-9 is also designated as PARC (p53-associated parkin-like cytoplasmic protein) [41]. CRL E3s consist of a scaffold protein Cullin, an adaptor protein, a substrate receptor protein, and/or a RING protein [41, 44].

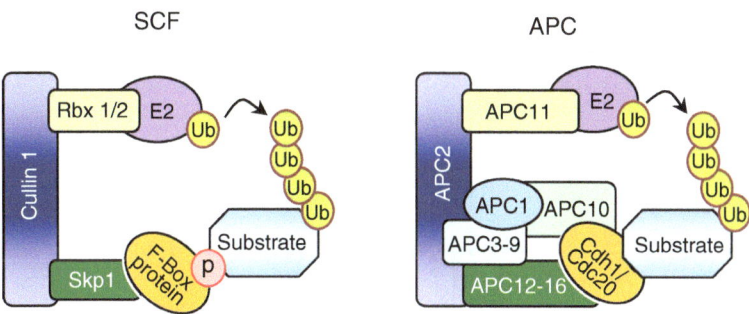

Fig. 1.2 A schematic illustration of similar structural organization of the multiple-subunit SCF and APC E3 ubiquitin ligases

Among the multi-subunit RING type E3 ligases, the SCF (Skp1-Cullin1-F-box) complex and the APC complex (Anaphase Promoting Complex/Cyclosome, also known as CRL-like ligase) have been extensively characterized [45] and are indispensable in regulating timely and precise UPS-dependent degradation of key Cyclins or Cdk kinase inhibitors (CdkIs) to drive cell cycle progression. SCF and APC complexes are evolutionarily related, as both of them contain a homologous Cullin protein (Cul1 in SCF and APC2 in APC) and a RING finger protein (Roc1/Rbx1 in SCF, and APC11 in APC) [5, 18] (Fig. 1.2). APC and SCF are the major driving forces to regulate cell cycle progression [45], as the SCF complex primarily regulates entry into S phase by degradation of Cdk kinase inhibitors and G1 Cyclins [45], whereas APC governs cell cycle progression through the G2 to M transition [2].

The SCF complex is composed of the scaffold protein Cullin1, the RING finger protein Rbx1 which recruits the E2 enzyme, and the adaptor protein Skp1 (S phase kinase associated protein 1) to bridge F-box proteins that mediates substrate recognition and recruitment, to the SCF complex [5, 46] (Figs. 1.2 and 1.3). The human genome encodes 69 putative F-box proteins that contain a homologous 40-amino-acid F-box domain initially identified in Cyclin F [47]. The F-box domain mediates the interaction between the F-box proteins and the Skp1 component of SCF. Each F-box protein also has substrate binding domains, and according to these domains F-box proteins are further divided into three major classes: the FBXW family (contains WD40 substrate binding domains), the FBXL family (contains LRR (leucine-rich repeats) substrate binding domains), and the FBXO family (contains other motifs such as kelch repeats or proline-rich motifs to bind substrates) (Fig. 1.3). Through individual F-box proteins, SCF can target a broad range of substrates for ubiquitination and destruction to exert its various physiological functions [3, 45].

The APC complex is composed of 14 subunits [48], which together forms a general structure that is functionally similar to that of SCF ligases (Fig. 1.2). Specifically, the APC core complex consists of three subcomplexes: a scaffolding subunit, a catalytic subunit, and a tetratricopeptide repeat (TPR) arm. It has been revealed that

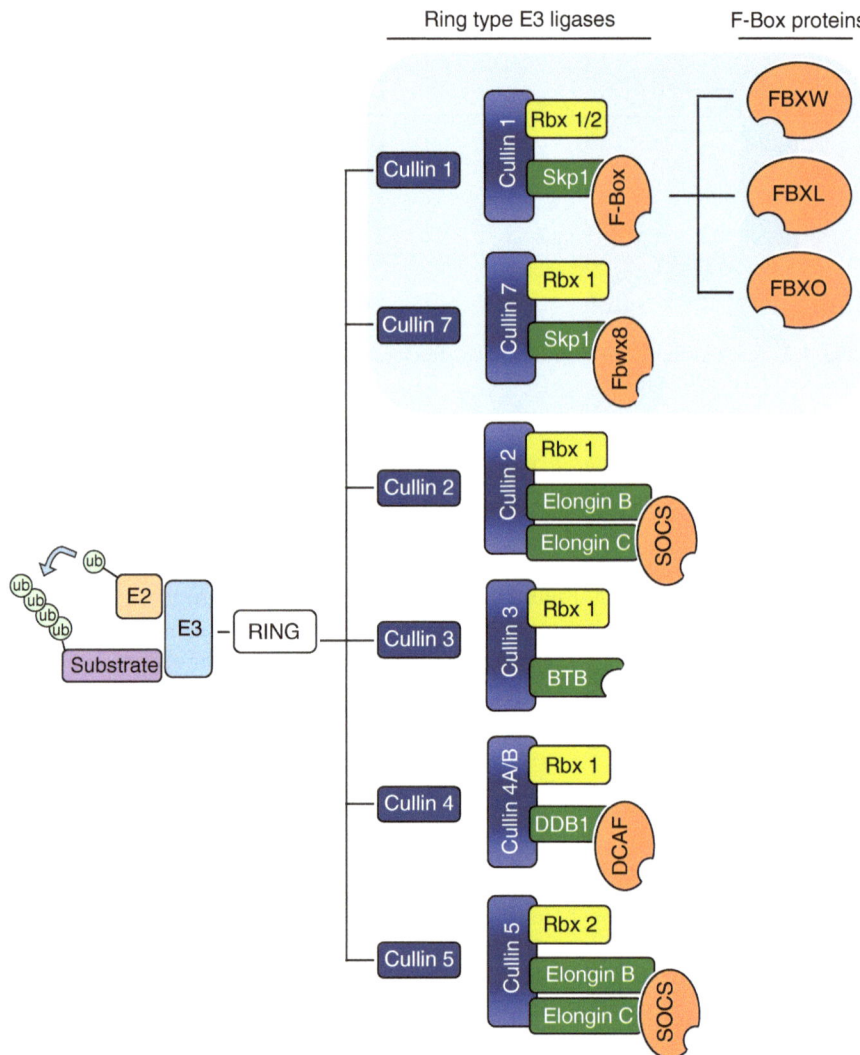

Fig. 1.3 A schematic illustration of various types of Cullin-based E3 ubiquitin ligases

APC is activated by association with the substrate-recruiting module, either Cdc20 [6–8] (cell division cycle 20 homologue) or Cdh1 [9, 10, 49]. These two adaptor proteins control APC to interact with its specific targets for degradation by the 26S proteasome at different times during the cell cycle progression, leading to refined cell cycle regulation. Specifically, the APC core complex binds the substrate adaptor Cdc20 to form APC^{Cdc20} to orchestrate the initiation of anaphase and exit from mitosis, and the substrate adaptor Cdh1 to form APC^{Cdh1} to govern the transition through late mitosis to the S phase. The core complex and co-activators assemble into an approximately 1.5 MDa holoenzyme complex. Recent structural studies have revealed the general architecture of APC, and have also shed light on the molecular mechanism by which APC recognizes substrates and mediates their ubiquitination [50–53].

1.4 Substrate Recognition by SCF and APC

A key difference between APC and SCF lies within their modes in recognizing their substrates. APC typically recognizes protein sequence (or degrons) present in substrates that tend to be unmodified. Specifically, APC^{Cdc20} typically degrades proteins with D-boxes [54], while APC^{Cdh1} targets a wide range of substrates containing the D-box [54], KEN-box [55], A-box [56, 57], O-box [58], CRY box [59], or GxEN box [60]. On the other hand, although F-box proteins also tend to recognize specific degron sequences within target substrates [45], prior modifications of these degrons are typically required to trigger the interaction between the F-box proteins and their substrate proteins, such as phosphorylation, methylation, acetylation and glycosylation (Fig. 1.4) [61]. For example, the consensus sequence D-pS-G-X-X-pS (X represents any amino acid) degron is recognized by β-TRCP where phosphorylation of both serine residues by specific kinases is necessary for β-TRCP-mediated ubiquitination and degradation [45]. Similarly, FBW7 substrates typically contain a conserved CPD (Cdc4 phospho-degron) sequence (L)-X-pT/pS-P-(P)-X-pS/pT/E/D (X represents any amino acid) [1]. Phosphorylation of the substrate within its degron is also required for FBW7 to recognize and target its substrate for subsequent ubiquitination. It is further demonstrated that phospho-degrons could be phosphorylated by a single kinase or multiple kinases, adding an additional layer of regulation in controlling substrate recognition and degradation [61, 62].

Beyond phosphorylation, some F-box proteins also recognize degrons with glycosylation (Fig. 1.4). For instance, FBXO6 binds the glycosylated degron in TCR (T cell receptor α-chain). Other than FBXO6, FBXO2 can bind proteins attached to N-linked high-mannose oligosaccharides and subsequently leads to polyubiquitination of N-glycosylated proteins such as pre-integrin β1 [63]. Interestingly, recent studies revealed that the F-box protein Cyclin F recognizes an unmodified Arg-X-Leu degron motif for the degradation of substrates such as centrosomal protein CP110 [64] and ribonucleotide reductase subunit M2 (RRM2) [64]. These results suggest that other mechanisms including the restriction of degron access, regulation of F-box protein localization, or F-box protein stability might be involved in control degradation mediated by Cyclin F [61]. With no doubt, further in-depth investigations are required to more completely understand the regulatory mechanisms governing F-box substrate recognition.

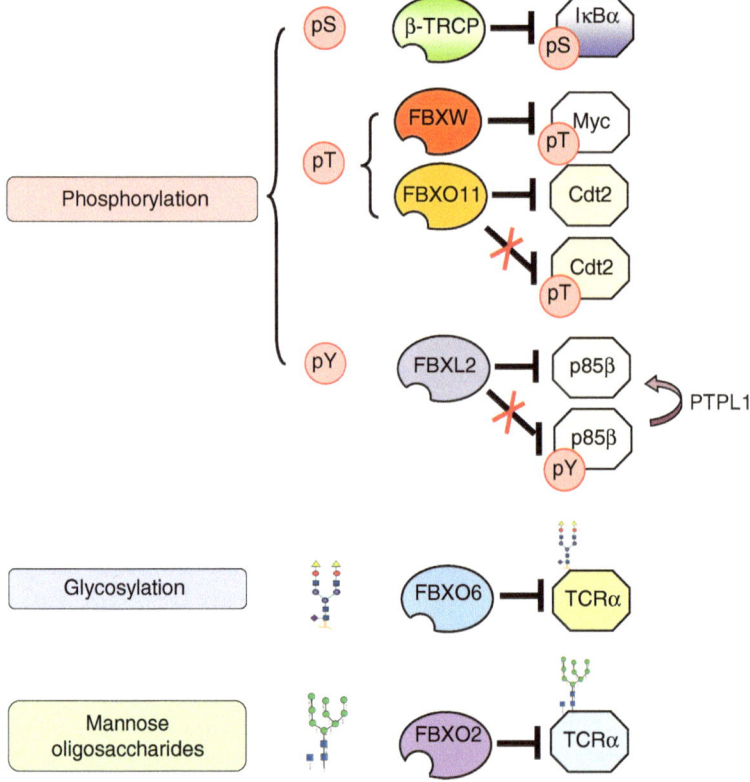

Fig. 1.4 An illustration of various modes of substrate recognition that is regulated by different types of substrate modifications

1.5 Interplays Between SCF and APC

Various cell cycle progression events regulated by SCF and APC are also precisely coordinated by interplays between these two major E3 ligase complexes (Fig. 1.5). For mitotic entry, SCF$^{β\text{-TRCP}}$ promotes the degradation of the Cdc20 inhibitor Emi1 at early M phase [65, 66]. Moreover, SCF$^{β\text{-TRCP}}$ reciprocally controls APC^{Cdh1} activity by promoting Cdh1 ubiquitination and its subsequent degradation [67]. Furthermore, two Cdh1 substrates Cyclin A and Plk1, were identified to promote Cdh1 phosphorylation, thereby triggering its ubiquitination and subsequent degradation by SCF$^{β\text{-TRCP}}$ [67]. Concomitantly, increased activity of SCF^{Skp2} leads to the degradation of intrinsic Cdk2 inhibitors p27 and p21 to induce the activation of the Cdk2/Cyclin A, which facilitates Cdh1 phosphorylation and the subsequent dissociation of Cdh1 from the APC complex to further inactivate APC^{Cdh1} activity

Fig. 1.5 Interplays between the SCF and APC E3 ubiquitin ligases

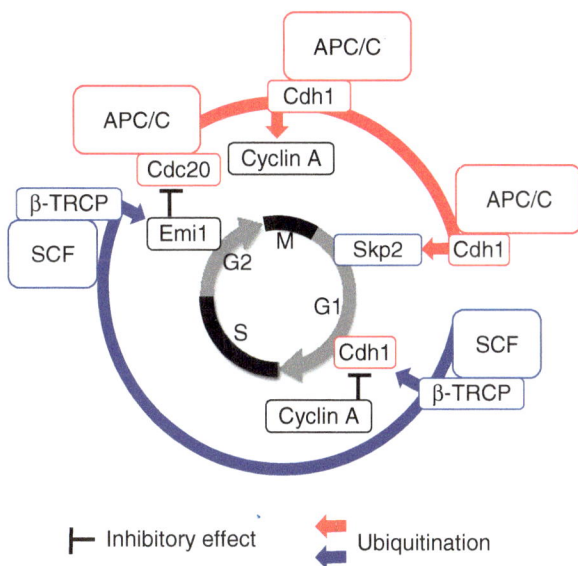

[68, 69]. On the other hand, APC^Cdh1 suppresses the activity of SCF by earmarking Skp2 for ubiquitination and destruction in early G1 phase [70]. These studies reveal that one E3 ubiquitin ligase can target other E3 ubiquitin ligase(s) for ubiquitination and subsequent degradation, thereby suggesting the existence of an "E3 ligase cascade", in analogy to the well-characterized "kinase cascade" (Fig. 1.6a, b). These findings also illustrate an elegant dual repression system among E3 ligase complexes to control timely cell cycle transitions through inter-regulation of SCF and APC activities [67]. Furthermore, accumulating evidence also suggested that extensive crosstalks exist between F-box E3 ligases and kinases [45, 61] (Fig. 1.6c), yet further work is necessary to continue to elucidate the exact molecular mechanisms of these F-box E3 ligases/kinases signaling pathways and their biological relevance.

Numerous intriguing studies have highlighted that disruption of signaling pathways by aberrant regulation of SCF and APC ubiquitin ligases causes abnormal cell cycle regulation, which in turn leads to tumorigenesis. In this book, we discuss the physiological roles of SCF and APC activities in normal cells and their roles in regulating tumorigenesis. We also review key aspects of SCF and APC mediated ubiquitination and proteolysis processes in both normal and dysregulated cell cycles to reveal how aberrant activities of SCF and APC contribute to tumorigenesis. Moreover, we finally discuss possible clinical implications based on targeting SCF and APC functions or their downstream substrates as therapeutic interventions for a variety of human diseases including cancer.

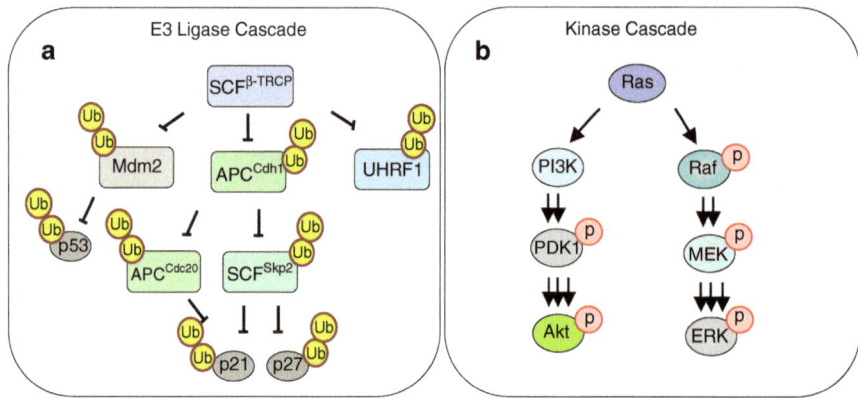

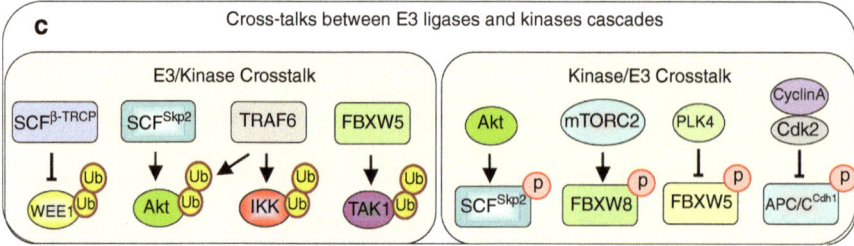

Fig. 1.6 A schematic illustration of a representative F-Box E3 ligase signal cascade and crosstalks between F-Box E3 ligases and kinases

References

1. Crusio KM, et al. The ubiquitous nature of cancer: the role of the SCF(Fbw7) complex in development and transformation. Oncogene. 2010;29(35):4865–73.
2. Nakayama KI, Nakayama K. Ubiquitin ligases: cell-cycle control and cancer. Nat Rev Cancer. 2006;6(5):369–81.
3. Welcker M, Clurman BE. FBW7 ubiquitin ligase: a tumour suppressor at the crossroads of cell division, growth and differentiation. Nat Rev Cancer. 2008;8(2):83–93.
4. Cardozo T, Pagano M. The SCF ubiquitin ligase: insights into a molecular machine. Nat Rev Mol Cell Biol. 2004;5(9):739–51.
5. Peters JM. The anaphase promoting complex/cyclosome: a machine designed to destroy. Nat Rev Mol Cell Biol. 2006;7(9):644–56.
6. Li M, York JP, Zhang P. Loss of Cdc20 causes a securin-dependent metaphase arrest in two-cell mouse embryos. Mol Cell Biol. 2007;27(9):3481–8.
7. Manchado E, et al. Targeting mitotic exit leads to tumor regression in vivo: modulation by Cdk1, Mastl, and the PP2A/B55alpha, delta phosphatase. Cancer Cell. 2010;18(6):641–54.
8. Yin S, et al. Cdc20 is required for the anaphase onset of the first meiosis but not the second meiosis in mouse oocytes. Cell Cycle. 2007;6(23):2990–2.
9. Garcia-Higuera I, et al. Genomic stability and tumour suppression by the APC/C cofactor Cdh1. Nat Cell Biol. 2008;10(7):802–11.
10. Li M, et al. The adaptor protein of the anaphase promoting complex Cdh1 is essential in maintaining replicative lifespan and in learning and memory. Nat Cell Biol. 2008;10(9):1083–9.

11. Malumbres M, Barbacid M. Cell cycle, CDKs and cancer: a changing paradigm. Nat Rev Cancer. 2009;9(3):153–66.
12. Besson A, Dowdy SF, Roberts JM. CDK inhibitors: cell cycle regulators and beyond. Dev Cell. 2008;14(2):159–69.
13. Resnitzky D, et al. Acceleration of the G1/S phase transition by expression of cyclins D1 and E with an inducible system. Mol Cell Biol. 1994;14(3):1669–79.
14. Nishimoto T, Uzawa S, Schlegel R. Mitotic checkpoints. Curr Opin Cell Biol. 1992;4(2): 174–9.
15. Wolowiec D, Ffrench M. Cyclins A and B: redundancy and specificity. Pathol Biol (Paris). 1993;41(6):547–53.
16. Morgan DO. Principles of CDK regulation. Nature. 1995;374(6518):131–4.
17. Olashaw N, Pledger WJ. Paradigms of growth control: relation to Cdk activation. Sci STKE. 2002;2002(134):re7.
18. Dai Y, Grant S. Cyclin-dependent kinase inhibitors. Curr Opin Pharmacol. 2003;3(4): 362–70.
19. Komander D, Rape M. The ubiquitin code. Annu Rev Biochem. 2012;81:203–29.
20. Hershko A, Ciechanover A. The ubiquitin system. Annu Rev Biochem. 1998;67:425–79.
21. Varshavsky A. The ubiquitin system, an immense realm. Annu Rev Biochem. 2012;81: 167–76.
22. Eldridge AG, O'Brien T. Therapeutic strategies within the ubiquitin proteasome system. Cell Death Differ. 2010;17(1):4–13.
23. Hoeller D, Dikic I. Targeting the ubiquitin system in cancer therapy. Nature. 2009;458(7237): 438–44.
24. Nalepa G, Rolfe M, Harper JW. Drug discovery in the ubiquitin-proteasome system. Nat Rev Drug Discov. 2006;5(7):596–613.
25. Pickart CM. Mechanisms underlying ubiquitination. Annu Rev Biochem. 2001;70:503–33.
26. Kulathu Y, Komander D. Atypical ubiquitylation—the unexplored world of polyubiquitin beyond Lys48 and Lys63 linkages. Nat Rev Mol Cell Biol. 2012;13(8):508–23.
27. Baboshina OV, Haas AL. Novel multiubiquitin chain linkages catalyzed by the conjugating enzymes E2EPF and RAD6 are recognized by 26 S proteasome subunit 5. J Biol Chem. 1996;271(5):2823–31.
28. Schwarz LA, Patrick GN. Ubiquitin-dependent endocytosis, trafficking and turnover of neuronal membrane proteins. Mol Cell Neurosci. 2012;49(3):387–93.
29. Ulrich HD, Walden H. Ubiquitin signalling in DNA replication and repair. Nat Rev Mol Cell Biol. 2010;11(7):479–89.
30. Acconcia F, Sigismund S, Polo S. Ubiquitin in trafficking: the network at work. Exp Cell Res. 2009;315(9):1610–8.
31. Clague MJ, Coulson JM, Urbe S. Cellular functions of the DUBs. J Cell Sci. 2012;125(Pt 2): 277–86.
32. Burrows JF, Johnston JA. Regulation of cellular responses by deubiquitinating enzymes: an update. Front Biosci. 2012;17:1184–200.
33. Fraile JM, et al. Deubiquitinases in cancer: new functions and therapeutic options. Oncogene. 2012;31(19):2373–88.
34. DeSalle LM, Pagano M. Regulation of the G1 to S transition by the ubiquitin pathway. FEBS Lett. 2001;490(3):179–89.
35. Li W, et al. Genome-wide and functional annotation of human E3 ubiquitin ligases identifies MULAN, a mitochondrial E3 that regulates the organelle's dynamics and signaling. PLoS One. 2008;3(1):e1487.
36. Bedford L, et al. Ubiquitin-like protein conjugation and the ubiquitin-proteasome system as drug targets. Nat Rev Drug Discov. 2011;10(1):29–46.
37. Deshaies RJ, Joazeiro CA. RING domain E3 ubiquitin ligases. Annu Rev Biochem. 2009;78:399–434.
38. Petroski MD, Deshaies RJ. Function and regulation of cullin-RING ubiquitin ligases. Nat Rev Mol Cell Biol. 2005;6(1):9–20.

39. Skaar JR, Pagano M. Control of cell growth by the SCF and APC/C ubiquitin ligases. Curr Opin Cell Biol. 2009;21(6):816–24.
40. Rotin D, Kumar S. Physiological functions of the HECT family of ubiquitin ligases. Nat Rev Mol Cell Biol. 2009;10(6):398–409.
41. Metzger MB, Hristova VA, Weissman AM. HECT and RING finger families of E3 ubiquitin ligases at a glance. J Cell Sci. 2012;125(Pt 3):531–7.
42. Hua Z, Vierstra RD. The cullin-RING ubiquitin-protein ligases. Annu Rev Plant Biol. 2011;62:299–334.
43. Sarikas A, Hartmann T, Pan ZQ. The cullin protein family. Genome Biol. 2011;12(4):220.
44. Duda DM, et al. Structural regulation of cullin-RING ubiquitin ligase complexes. Curr Opin Struct Biol. 2011;21(2):257–64.
45. Frescas D, Pagano M. Deregulated proteolysis by the F-box proteins SKP2 and beta-TrCP: tipping the scales of cancer. Nat Rev Cancer. 2008;8(6):438–49.
46. Zheng N, et al. Structure of the Cul1-Rbx1-Skp1-F boxSkp2 SCF ubiquitin ligase complex. Nature. 2002;416(6882):703–9.
47. Bai C, et al. SKP1 connects cell cycle regulators to the ubiquitin proteolysis machinery through a novel motif, the F-box. Cell. 1996;86(2):263–74.
48. Lipkowitz S, Weissman AM. RINGs of good and evil: RING finger ubiquitin ligases at the crossroads of tumour suppression and oncogenesis. Nat Rev Cancer. 2011;11(9):629–43.
49. McLean JR, et al. State of the APC/C: organization, function, and structure. Crit Rev Biochem Mol Biol. 2011;46(2):118–36.
50. Schreiber A, et al. Structural basis for the subunit assembly of the anaphase-promoting complex. Nature. 2011;470(7333):227–32.
51. da Fonseca PC, et al. Structures of APC/C(Cdh1) with substrates identify Cdh1 and Apc10 as the D-box co-receptor. Nature. 2011;470(7333):274–8.
52. Buschhorn BA, et al. Substrate binding on the APC/C occurs between the coactivator Cdh1 and the processivity factor Doc1. Nat Struct Mol Biol. 2011;18(1):6–13.
53. Jin L, et al. Mechanism of ubiquitin-chain formation by the human anaphase-promoting complex. Cell. 2008;133(4):653–65.
54. Glotzer M, Murray AW, Kirschner MW. Cyclin is degraded by the ubiquitin pathway. Nature. 1991;349(6305):132–8.
55. Pfleger CM, Kirschner MW. The KEN box: an APC recognition signal distinct from the D box targeted by Cdh1. Genes Dev. 2000;14(6):655–65.
56. Castro A, et al. The D-Box-activating domain (DAD) is a new proteolysis signal that stimulates the silent D-Box sequence of Aurora-A. EMBO Rep. 2002;3(12):1209–14.
57. Littlepage LE, Ruderman JV. Identification of a new APC/C recognition domain, the A box, which is required for the Cdh1-dependent destruction of the kinase Aurora-A during mitotic exit. Genes Dev. 2002;16(17):2274–85.
58. Araki M, et al. Degradation of origin recognition complex large subunit by the anaphase-promoting complex in Drosophila. EMBO J. 2003;22(22):6115–26.
59. Reis A, et al. The CRY box: a second APCcdh1-dependent degron in mammalian cdc20. EMBO Rep. 2006;7(10):1040–5.
60. Castro A, et al. Xkid is degraded in a D-box, KEN-box, and A-box-independent pathway. Mol Cell Biol. 2003;23(12):4126–38.
61. Skaar JR, Pagan JK, Pagano M. Mechanisms and function of substrate recruitment by F-box proteins. Nat Rev Mol Cell Biol. 2013;14(6):369–81.
62. Hao B, et al. Structure of a Fbw7-Skp1-cyclin E complex: multisite-phosphorylated substrate recognition by SCF ubiquitin ligases. Mol Cell. 2007;26(1):131–43.
63. Yoshida Y, et al. E3 ubiquitin ligase that recognizes sugar chains. Nature. 2002;418(6896): 438–42.
64. D'Angiolella V, et al. SCF(Cyclin F) controls centrosome homeostasis and mitotic fidelity through CP110 degradation. Nature. 2010;466(7302):138–42.

65. Margottin-Goguet F, et al. Prophase destruction of Emi1 by the SCF(betaTrCP/Slimb) ubiquitin ligase activates the anaphase promoting complex to allow progression beyond prometaphase. Dev Cell. 2003;4(6):813–26.
66. Guardavaccaro D, et al. Control of meiotic and mitotic progression by the F box protein beta-Trcp1 in vivo. Dev Cell. 2003;4(6):799–812.
67. Fukushima H, et al. SCF-mediated Cdh1 degradation defines a negative feedback system that coordinates cell-cycle progression. Cell Rep. 2013;4(4):803–16.
68. Lukas C, et al. Accumulation of cyclin B1 requires E2F and cyclin-A-dependent rearrangement of the anaphase-promoting complex. Nature. 1999;401(6755):815–8.
69. Sorensen CS, et al. A conserved cyclin-binding domain determines functional interplay between anaphase-promoting complex-Cdh1 and cyclin A-Cdk2 during cell cycle progression. Mol Cell Biol. 2001;21(11):3692–703.
70. Wei W, et al. Degradation of the SCF component Skp2 in cell-cycle phase G1 by the anaphase-promoting complex. Nature. 2004;428(6979):194–8.

Chapter 2
The Role of FBXW Subfamily of F-box Proteins in Tumorigenesis

Alan W. Lau, Yueyong Liu, Adriana E. Tron, Hiroyuki Inuzuka, and Wenyi Wei

Abstract In the human genome, 69 putative F-box proteins have been identified which could form a variety of different SCF type E3 ligase complexes to specifically target a wide range of proteins for degradation. F-box proteins can be further subclassified into three families largely based on the presence of three recognizable domains: WD repeats, leucine-rich repeats (LRR), and other types of protein interaction domains. The FBXW subfamily comprises ten proteins that contain both the F-box motif and the WD40 repeat domain, including FBXW-1 (also known as beta-transducin repeat-containing protein and β-TRCP1), FBXW-2, FBXW-4, FBXW-5, FBXW-7, FBXW-8, FBXW-9, FBXW-10, FBXW-11 (also known as β-TRCP2), and FBXW-12 (Fig. 2.1).

Over the past 17 years, intensive research efforts have been devoted, using both mouse genetic models and biochemical approaches, to identify specific ubiquitin substrates for a given F-box protein, which have helped the scientific community to reveal the important contributions of F-box proteins including the three well-studied E3 ligases SCF^{Skp2}, SCF^{FBW7}, and SCF$^{\beta\text{-TRCP}}$ in human cancers (Welcker and Clurman, Nat Rev Cancer 8(2):83–93, 2008; Frescas and Pagano, Nat Rev Cancer 8(6):438–449, 2008). Mechanistically, misregulated degradation of oncoproteins or tumor suppressors by various SCF E3 ligases could lead to tumorigenesis. Thus, F-box proteins could function as either oncoproteins or tumor suppressors, depending on the functional outputs of their ubiquitin substrates.

In this chapter, we summarize the recent genetic, pathological, and biochemical evidence revealing a possible role for FBXW subfamily of F-box proteins in tumorigenesis. To this end, we focus our discussion on our current knowledge accumulated in three major categories: physiological evidence (genetically engineered

Alan W. Lau and Yueyong Liu have contributed equally to this chapter.

A.W. Lau • Y. Liu • A.E. Tron • H. Inuzuka (✉) • W. Wei (✉)
Department of Pathology, Beth Israel Deaconess Medical Center,
Harvard Medical School, Boston, MA 02215, USA
e-mail: hinuzuka@bidmc.harvard.edu; wwei2@bidmc.harvard.edu

H. Inuzuka and W. Wei, *SCF and APC E3 Ubiquitin Ligases in Tumorigenesis*, 15
SpringerBriefs in Cancer Research, DOI 10.1007/978-3-319-05026-3_2,
© The Author(s) 2014

animal models), pathological evidence (human clinical cancer relevance), and bio-chemical evidence (ubiquitin substrates), all combined will allow for a more thorough understanding for their roles in tumorigenesis (Table 2.4). Given the fact that physiological evidence (primarily through mouse modeling studies) is considered the strongest supportive evidence indicating an involvement of a given F-box protein in tumorigenesis (Tables 2.1 and 2.3), we limit our discussion to those FBXW members that have available mouse genetic models.

Keywords F-box • SCF • FBXW subfamily • FBXW7 • beta-TRCP1 • beta-TRCP2 • FBXW8 • Cullin 1 • Cullin 7 • Tumor suppressor • Oncoprotein • Mouse model • Physiological function

2.1   FBXW1 and FBXW11 (β-TRCP1 and β-TRCP2) have Context-Dependent Functions in Cancer

2.1.1   The SCF$^{β\text{-}TRCP}$ E3 Ubiquitin Ligase

FBXW1, also termed β-TRCP (beta-transducin repeat containing protein), is conserved across species and has been well characterized in *Homo sapiens* (β-TRCP1 and β-TRCP2), *Drosophila melanogaster* (Slimb), and *Xenopus* (β-TRCP) [2]. Similar to other F-box proteins, β-TRCP generally recognizes a phosphorylated consensus sequence (DpSGxxpS) in its target substrates [3]. Structurally, β-TRCP contains an F-box motif at its N-terminus, multiple WD-40 repeats at its C-terminus and a dimerization domain near the F-box motif [2, 4]. The F-box motif is a critical 40-amino-acid sequence required for interaction with Skp1, an adaptor component of the SCF complex [5]. WD-40 repeats are protein–protein interaction motifs that form direct contact with target substrates [6]. At this point, the importance of the dimerization domain is currently unknown due to lack of genetic evidence. While it has been previously reported that the homo-dimerization and hetero-dimerization between β-TRCP1 and β-TRCP2 might be important for substrate specificity [4], additional studies are still necessary to further understand the role of dimerization in β-TRCP function (Fig. 2.1).

In humans, β-TRCP exists as two homologues, β-TRCP1 and β-TRCP2, which are encoded by two distinct genes. Structurally, both isoforms contain an F-box domain and all seven WD-40 repeats, with only noticeable sequence differences in their N-terminal regions [7]. To date, the differences between these two proteins remain unknown. Recent studies have suggested that β-TRCP1 and β-TRCP2 are functionally redundant, at least based on various in vitro biochemical assays. If so, this might explain why mice lacking β-TRCP1 develop normally with only minor spermatogenesis defects, as β-TRCP2 is still present and may compensate for loss of β-TRCP1 [8] (Table 2.1). Furthermore, in mouse embryonic fibroblasts, β-TRCP1,

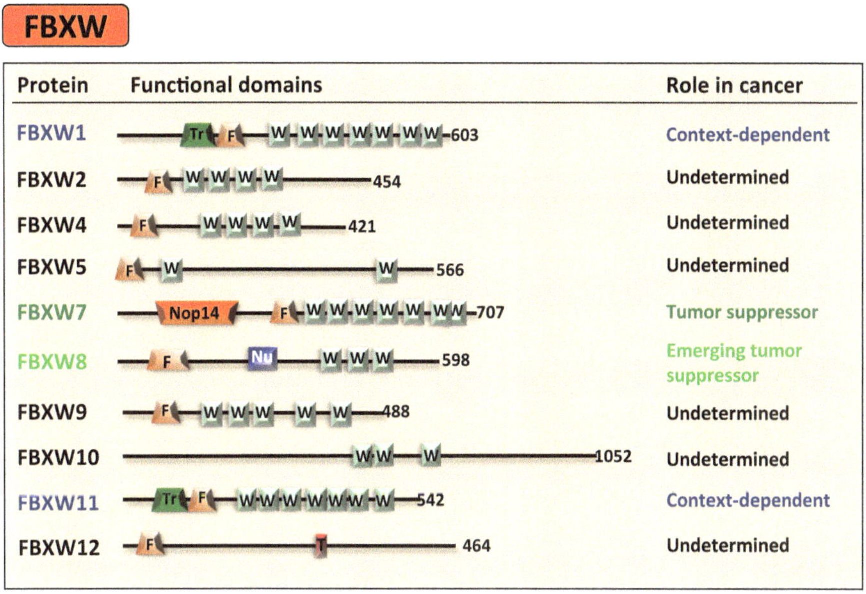

FBXW

Protein	Functional domains	Role in cancer
FBXW1	——[Tr]-[F]—W-W-W-W-W-W-W—603	Context-dependent
FBXW2	—[F]—W-W-W-W———454	Undetermined
FBXW4	—[F]—W-W-W-W———421	Undetermined
FBXW5	[F]—W——————W—566	Undetermined
FBXW7	——[Nop14]——[F]—W-W-W-W-W-W-W—707	Tumor suppressor
FBXW8	—[F]———[Nu]———W-W-W———598	Emerging tumor suppressor
FBXW9	—[F]—W-W-W—W—W——488	Undetermined
FBXW10	————————————W-W—W————1052	Undetermined
FBXW11	——[Tr]-[F]—W-W-W-W-W-W-W—542	Context-dependent
FBXW12	—[F]————[T]———— 464	Undetermined

Fig. 2.1 A schematic illustration of functional domains of all known FBXW proteins and their possible roles in cancer. *F* F-box motif, *W* WD40 repeat, *Tr* D domain of beta-TRCP, *Nop14* Nop14-like family domain, *Nu* Nucleoporin Nup120/160-like domain

while contributes, is not completely required for the degradation of certain substrates [8]. Obviously, genetically modified mice deleted of β-TRCP1 and/or β-TRCP2 would be required to fully understand the role of β-TRCP in various cellular processes and whether the two homologues play an individual, synergistic, or redundant role under different experimental conditions.

β-TRCP1/2 targets a range of substrates for ubiquitination and subsequent degradation by the 26S proteasome (Table 2.2). As such, β-TRCP plays a vital regulatory role in multiple biological processes. Interestingly, many known substrates of β-TRCP1/2 are involved in various stages of cancer development and progression. For instance, some known substrates include cell cycle regulators Emi1 [9], Cdc25A [10, 11], Wee1A [12], Cyclin D1 [13], and BTG [14]. Consistent with a role in cell cycle regulation, *Btrc1*$^{-/-}$ mouse fibroblasts displayed increased genetic instability [9, 12] while Drosophila *Slimb* mutants exhibited centrosome and mitotic defects [15]. Furthermore, β-TRCP1/2 has also been shown to ubiquitinate proteins involved in cell migration. These substrates include the transcription factor Snail [16] as well as the extracellular matrix protein fibronectin [17]. Moreover, targets of β-TRCP have also been shown to regulate the cellular expression of various apoptotic proteins including Mcl-1 [18], BimEL [19], PDCD4 [20], and Pro-caspase-3 [21].

Table 2.1 Summary of the knockout mouse models for the FBXW class of F-Box proteins

F-box protein	Whole-body knockout/ phenotype	Tissue-specific knockout	
		Tissue/cells	Phenotype
FBXW1 (β-TRCP1) FBXW11 (β-TRCP2) Context-dependent	*Btrc1*$^{-/-}$: Viable and fertile (male) [8] Reduced fertility (male) [130] Hypoplastic phenotype in mammary glands (female) [26] Abnormal retinal development [139]		
FBXW7 Tumor suppressor	*Fbxw7*$^{-/-}$: embryonic lethal [40, 41] *Fbxw7*$^{+/-}$: radiation induced tumorigenesis [54] *Fbxw7β*$^{-/-}$: normal [140]	Gut Cerebellar anlage	Intestinal tumor [115] Decreased cerebellar size and defects in axonal arborization [141]
	Compound mice models *Fbxw7*$^{-/-}$*p53*$^{-/-}$: colorectal cancer (CRC) [59]	Brain Liver	Die after birth [82, 142] Hepatomegaly and steatohepatitis [117]
	Fbxw7$^{+/-}$*Apc*$^{+/-}$: impaired intestinal progenitor and neural stem cell differentiation [143]	Intestine Hematopoietic stem cells (HSCs)	Impaired differentiation of progenitor cells [48] Defective maintenance of quiescence [113]
	Fbxw7$^{+/-}$*p53*$^{+/-}$: tumors in epithelial tissues [54]	Hematopoietic tissue T cell	T-ALL [134] Thymic lymphoma [112]
	Fbxw7$^{+/-}$*Pten*$^{-/-}$: accelerated tumor formation [144]	Leukemia initiating cell	Abrogation of quiescence [122] Tumor inhibition [123]
FBXW8 Emerging tumor suppressor	*Fbxw8*$^{-/-}$: smaller littermates [127] Abnormal placenta [126]		

2.1.2 The Oncogenic Role of β-TRCP in Cancer Development

As ubiquitin targets of β-TRCP play such indispensable roles in biological pathways related to the various stages of tumorigenesis, it is not surprising that dysregulation of SCF$^{β\text{-TRCP}}$ has been implicated in cancer development and progression. To date, multiple studies, both in vitro and in vivo, have suggested that β-TRCP might function as an oncogene and can promote cancer formation when overexpressed (Table 2.3). Currently, elevated levels of β-TRCP have been observed in colorectal cancer [22], hepatoblastoma [23], pancreatic cancer [24], and melanoma [25].

Table 2.2 Summary of identified ubiquitin substrates for β-TRCP1/2

Substrates	Functions/signaling pathways	Phospho-degron (phosphorylation sites)	References
β-catenin	Transcription factor, Wnt signaling pathway	DSGIHS	[6, 35, 147]
Gli2	Transcription factor, Shh signaling pathway	SSAYTVS, DSYDPIS	[148]
Gli3	Transcription factor, Shh signaling pathway	Multiple-degrons	[149, 150]
YAP	Transcription regulator, Hippo signaling pathway	DSGLSMS	[151]
TAZ	Transcription factor, Hippo signaling pathway	SREQSTDSG	[152]
HSF1	Transcription factor, Heat shock response	DSGSAHS	[153]
Nrf1	Transcription factor, Oxidative stress	DSGLS	[154]
Nrf2	Transcription factor, Oxidative stress	DSGIS, DSAPGS, DSGISL	[155]
ATF4	Transcription factor, Amino acid metabolic processes	DSGICMS	[156]
ESE-1	Transcription factor, Epithelial cell differentiation	SSHSSDSG	[157]
FOXO3a	Transcription factor, Apoptosis and metabolic processes	DSLSHS, DSLIST	[28]
c-Myc	Transcription factor, Cell proliferation	ESGSPS	[158]
p63γ	Transcription factor, Apoptosis	–	[159]
p53	Transcription factor, Apoptosis and cell cycle	GSRAHS	[36]
REST	Transcription factor, Neuronal differentiation and cell proliferation	SEGSDDSG, DSGIHS	[160, 161]
Smad4	Transcription factor, TGFβ signaling pathway	–	[162]
Snail	Transcription factor, Cell migration and Mesoderm formation	DSGKGS	[16]
STAT1	Transcription factor, Cytokine signaling pathway	–	[163]
Sp1	Transcription factor, Cell proliferation, apoptosis, differentiation, and lipid metabolism	DSGAGS	[164, 165]
Twist	Transcription factor, Cell migration and myogenesis	DSLSNS	[166]
BTG	Transcription factor, Cell cycle	–	[14]
IkBα	Inhibitor of NFκB signaling pathway	DSGLDS	[7, 34, 35]
IkBβ	Inhibitor of NFκB signaling pathway	DSGLGSS	[33, 167]
IkBε	Inhibitor of NFκB signaling pathway	DGSIGS	[33]
p105	Transcription factor, NFκB signaling pathway	DSGVETS	[168–170]
p100	Transcription factor, NFκB signaling pathway	DSAYGS	[171–173]

(continued)

Table 2.2 (continued)

Substrates	Functions/signaling pathways	Phospho-degron (phosphorylation sites)	References
Bcl10	Adaptor protein, NFκB signaling pathway	DTLVES	[174]
CIKS	Adaptor protein, NFκB signaling pathway	–	[175]
IRAK1	Protein kinase, NFkB signaling pathway	–	[176]
Cdc25A	Protein phosphatase, Cell cycle	SSESTDSG, DDGFLD	[10, 11]
Cdc25B	Protein phosphatase, Cell cycle	DDGFVD, DSGFCLDS, DAGLCMDSPSP	[177, 178]
Plk4	Protein kinase, Cell cycle	DSGHAT	[179, 180]
Wee1A	Protein kinase, Cell cycle	DSAFQE, EEGFGSSS	[12]
Bora	Aurora A kinase activator, Cell cycle	DSGYNT	[181]
Emi1	Inhibitor of APC/C, Cell cycle	DSGYSS	[9]
Cyclin D1	Cyclin, Cell cycle	EEVDLACT	[13]
Securin	Regulatory protein, Cell cycle	DDAYPE	[182]
Per1	Component of circadian core oscillator, Circadian rhythm	TSGCSS	[183]
Per2	Component of circadian core oscillator, Circadian rhythm	SSGYGS	[184]
Claspin	Adaptor protein, DNA damage response	DSGQGS	[185–187]
FANCM	DNA helicases, DNA damage response	DSGYNS	[188]
RASSF1C	DAXX interacting protein, DNA damage response	SSGYCS	[189]
Pro-caspase-3	Cysteine–aspartic acid protease family, Apoptosis	–	[21]
BimEL	Bcl-2 family protein, Apoptosis	SSGYFSD	[19]
Mcl-1	Bcl2-family protein, Cell survival	STDGSLPST	[18]
PDCD4	RNA-binding protein, Apoptosis	DSGRGDS	[20]
AUF	DNA/RNA-binding protein, Transcriptional regulation	SSPRHS	[190]
CPEB	RNA-binding protein, mRNA regulation	TSGFSS	[191]
HuR	RNA-binding protein, mRNA stabilization	EEAMAIAS	[192]
TRF1	DNA-binding protein, Regulation of telomere length	–	[193]
BMI1	Component of a Polycomb group multiprotein PRC1-like complex, Transcriptional repression	DSGSDKANS	[194]
LPCAT1	Acyltransferase, Lipid metabolism	SDQDS	[195]
PFKFB3	Synthesis and degradation of fructose 2,6-bisphosphate, Metabolic processes	DSGLSS	[196]
eEF2K	Protein kinase, Protein synthesis	DSGYPS	[197, 198]
EPO-R	Receptor, Cytokine signaling pathway	DSGISTD	[199]

(continued)

Table 2.2 (continued)

Substrates	Functions/signaling pathways	Phospho-degron (phosphorylation sites)	References
GH-R	Receptor, Growth hormone signaling pathway	DSGRTS, DDSWVEFIE	[200]
IFNA-R1	Receptor, Interferon signaling pathway	DSGNYS	[201, 202]
IL-10R1	Receptor, Cytokine signaling pathway	DSGFGS, DSGICLQE	[203]
PRL-R	Receptor, Growth hormone signaling pathway	DSGRGS	[204]
VEGFR2	Receptor protein kinase, Angiogenesis	DSGLSLPT, DSGMYLAS, DSGTTLS, DDTDTT	[205, 206]
DEPTOR	mTOR inhibitor, PI3K signaling pathway	SSGYFSSS	[207–209]
PHLPP1	Protein phosphatase, PI3K/Akt signaling pathway	QSVLLT, DSLSVE	[210]
H-Ras	GTPase protein, Growth factor signaling pathway	–	[99]
Fibronectin	Extracellular matrix protein, Cell adhesion and migration processes	DSGVVYS, DSGSIVVS	[17]
RCAN1	Calcineurin A binding protein, Calcineurin signaling pathway	–	[211]
SPAR	Rap GTPase activating protein, Postsynaptic regulation	DSGIDT	[212]
Cortactin	Cortical actin binding protein, Actin cytoskeleton organization	–	[213]
DLG	Scaffolding protein, Cell adhesion	DSGLPS	[214]
FGD1	Cdc42 guanine-nucleotide exchange factor, Actin cytoskeleton organization	DSGIDS	[215]
FGD3	Cdc42 guanine-nucleotide exchange factor, Actin cytoskeleton organization	DSGIDS	[216]
Mdm2	E3 ubiquitin ligase, p53 signaling pathway	Multiple-degrons	[217]
Cdh1	Component of APC/C E3 ligase complex, Ubiquitin conjugation pathway	SPDDGNDVS	[218]
UHRF1	E3 ubiquitin ligase, Chromatin regulation	SDTDSG	[219]
USP47	Deubiquitinase, Ubiquitin conjugation pathway	DSGTDS	[220]
BST-2	Type 2 integral membrane protein, Inhibition of virus infection	–	[221, 222]
CD4 (through Vpu)	Membrane protein, T cell activation	DSGNES (Vpu)	[3]
PC2 (through TAZ)	Calcium permeable cation channel, Intracellular calcium homeostasis	–	[223]

Table 2.3 Summary of the transgenic mouse models for the FBXW class of F-Box proteins

F-box protein	Transgenic mouse model/phenotype				
	Whole-body expression/phenotype		Tissue-specific expression		
	Transgene	Phenotype	Tissue	Transgene	Phenotype
FBXW1 (β-TRCP1)					
FBXW11 (β-TRCP2)	Inducible β-TRCP2 knockdown in $Btrc1^{-/-}$ mice	Severe testicular defect [131]	Epidermis	Inducible expression of dominant negative β-TRCP2	Decreased UVB-induced edema, hyperplasia, and inflammatory response [145]
Context-dependent			Intestine, liver, and kidney	Dominant-negative or full-length β-TRCP1	Tumorigenesis [27]
FBXW7 Tumor suppressor	$Fbxw7^{R482Q}$	Died perinatally [146]	Intestine	$Fbxw7^{flR482Q/+}/Apc^{1322T+}$	Intestinal tumor [116]
	Fbxw7 mutant alleles	T-ALL [114]			

While most of these studies were performed utilizing cell culture based models, multiple in vivo mouse models have also been generated verifying the oncogenic capabilities of β-TRCP (Table 2.3). For example, Kudo et al. observed that tissue-specific expression of human β-TRCP1 in mouse mammary epithelia displayed increased cellular proliferation [26]. Furthermore, of the mice studied, 38 % developed mammary, ovarian, and uterine carcinomas. In addition to these findings, the Besnard-Guerin group also explored the implications of β-TRCP1 overexpression in mouse intestine, liver, and kidney. They observed that 46 % (16/35) of mice with elevated expression of β-TRCP1 developed either intestinal adenomas or hepatic or urothelial tumors [27]. Moreover, in an orthotopic tumor mouse model, Tsai et al. observed that cells overexpressing β-TRCP1 displayed increased tumorigenic activity [28]. It is both these in vivo and in vitro studies that have lent support for β-TRCP1/2 functioning as an oncogene. However, this may not always be the case, as multiple studies have also suggested that β-TRCP may display tumor suppressive capabilities in certain cellular contexts.

2.1.3   β-TRCP, a Possible Tumor Suppressor?

While predominately reported to function as an oncogene, β-TRCP may also function as a tissue-specific tumor suppressor. To this end, while uncommon, mutations in β-TRCP1/2 have been observed in various types of carcinomas. In a gastric cancer cell line, a WD-40 substrate binding domain mutation (F462S) was identified by Saitoh et al. They suggested that this mutation could lead to stabilization of β-catenin and subsequent tumorigenesis through hyper-activation of the Wnt signaling pathway [29]. These findings were later supported by the identification and analysis of five *β-TRCP* missense mutations (A99V, H342Y, H425Y, C206Y, and G260E) found in gastric cancers. In tissues carrying these mutations, moderate to strong β-catenin expression was detected by immunohistochemistry [30]. In further support of a possible tumor suppressive role of β-TRCP, mutations in this F-box protein have also been observed in prostate cancer [31] as well as breast cancer [32].

Interestingly, while substrates of β-TRCP, such as beta-catenin, exhibit oncogenic properties, many β-TRCP substrates are also known tumor suppressors. For example, IκB, a negative regulator of the NFκB oncogene, is a well-documented β-TRCP substrate [7, 33–35]. Furthermore, the FOXO3 tumor suppressor has recently been shown to be a target of β-TRCP1 as well [28]. Moreover, Xia et al. identified p53 to be a novel target of β-TRCP-mediated ubiquitination [36]. Therefore, based on these facts, it is currently difficult to classify β-TRCP as a bona fide tumor suppressor or oncoprotein. It is possible that β-TRCP, while largely oncogenic in nature, may also function as a tumor suppressor in certain cellular contexts or in specific tissues. Hence, as with all E3 ubiquitin ligases, further identification of novel substrates would provide researchers a more genuine understanding of the complex role for β-TRCP in cancer development. The discovery of additional putative substrates would further shed light on whether β-TRCP is an oncogene, a tumor suppressor in certain cellular contexts, or both. Without a doubt, generation of additional mouse genetic models would also be required to validate the tissue-specific or cellular context-dependent contribution of β-TRCP in tumorigenesis.

2.2 FBW7 Functions Largely as a Tumor Suppressor

The first member of the *FBW7* gene family was originally identified in budding yeast as a regulator of cell cycle-related proteins and designated as Cdc4 [37]. Over the past few decades, FBW7 (also known as FBXW7, hCdc4, hAgo, and SEL10) has been extensively studied and observed to be involved in a number of pivotal biological processes including cell growth, proliferation, differentiation, and survival [1]. Notably, FBW7 is considered a tumor suppressor through its negative regulation of many oncogenic proteins including c-Myc [38], Cyclin E [39], Notch [40, 41], c-Jun [42, 43], Mcl-1 [44, 45], and mTOR [46] (Tables 2.4 and 2.5). Moreover, mutations and deletions in *FBW7* have been identified in a wide spectrum of human cancers including T-cell acute lymphoblastic leukemia [51], cholangiocarcinoma, gastrointestinal cancer [48], bladder cancer [49], colon cancer [50], and prostate cancer [51]. Furthermore, emerging evidence have also demonstrated that dysregulation of FBW7 function can drive tumor development in humans [52]. In addition, microRNAs (miRNAs), as well as alternatively spliced isoforms, have been found to be involved in controlling FBW7 expression in human cancer settings [53]. Therefore, FBW7 is a general tumor suppressor in many human cancers.

Although great strides have been made to identify various downstream ubiquitin targets of FBW7 (Table 2.5), relatively little is known about the upstream signaling pathways that control FBW7 stability and cellular function. Additionally, as a haploinsufficient tumor suppressor [48, 54], understanding FBW7 and its regulatory axis is crucial for providing further insights into the precise molecular mechanism of FBW7 associated tumor suppression, and to further guide the development of therapeutic strategies that target the FBW7 pathway. In this section, we will primarily focus on the regulatory network of FBW7, knockout (KO) and conditional KO mouse models for *FBW7*, and the role of FBW7 in human cancer development.

2.2.1 FBW7 *Gene and Isoforms*

The human *FBW7* gene is located on chromosome 4q32 and encodes three transcripts, α, β and γ, derived from alternative transcriptional start sites that differ in the first exon, resulting in the generation of three different protein isoforms [55]. Each transcript contains its own promoter, although how these transcripts are regulated remains largely unknown with the exception of *FBW7β*, which is regulated by p53 [1]. Recently, Liu et al. reported newly identified alternative splice isoforms for

Table 2.4 Summary of the roles of FBXW class of F-box protein in cancer

Potential role in cancer (tumor suppressor or oncogene)	F-box protein	Gene symbol	Physiological evidence (mouse models)		Pathological evidence in cancer		Biochemical evidence (major substrates)
			Knockout	Transgenic	Overexpression	Mutation/deletion	
Context-dependent	FBXW1 (β-TRCP1)	BTRC, FWD1, FBXW1A	Yes [130, 131]	Yes [26, 27]	Yes [132]	Yes [133]	β-catenin, Cdc25A, IκB, DEPTOR
	FBXW11 (β-TRCP2)	FBXW11, HOS, FBXW1B, BTRC2, FBX1B					
Tumor suppressor	FBXW7	FBXW7, FBXW6, CDC4, SEL-10, FBX30	Yes [41, 54, 134]	Yes [114]		Yes [135–137]	Cyclin E, c-Myc, c-Jun, Mcl-1
Emerging tumor suppressor	FBXW8	FBXW8, Fbx29, FBXO29, Fbw6	Yes [126, 127]			CUL7 mutations in 3-M syndrome [138]	IRS1, TBC1D3, Cyclin D1

Table 2.5 Summary of the identified ubiquitin substrates for the Fbxw7

Substrates	Functions/signaling pathways	Phospho-degron (phosphorylation sites)	References
Cyclin E	Cyclin, Cell cycle	Thr62, Thr380, Ser384	[38, 73, 224]
c-Myc	Transcription factor, Cell proliferation	Thr58	[39, 55]
c-Jun	Transcription factor, Cell proliferation	Thr239	[42, 43]
Notch1	Transcription factor, Notch signaling pathway	Thr2512	[47, 84]
SREBP	Transcription factor, Lipid homeostasis	Thr426, Thr430	[225]
c-Myb	Transcription factor, Cell proliferation	Thr572	[226, 227]
JunB	Transcription factor, Cell proliferation	Thr255, Ser259	[228]
KLF2	Transcription factor, T cell quiescence and migration	Thr173, Ser177, Thr244, Ser248	[229]
KLF5	Transcription factor, Adipoctye differentiation and lipid metabolism	Ser303, Thr324	[139, 230]
KLF13	Transcription factor, Negative regulator of cell proliferation and erythrocyte differentiation	Ser119	[8]
C/EBPα	Transcription factor, Cell differentiation and body weight homeostasis	Thr222	[114]
C/EBPδ	Transcription factor, Immune and inflammatory responses	Thr156, Ser160	[131]
p100	Transcription factor	Ser707	[41, 59, 144]
HIF1α	Transcription factor, Hypoxia signaling pathway	Thr497	[116, 141]
Nrf	Transcription factor, Stress response	Ser271, Ser352	[143]
p63	Transcription factor, DNA damage response and cell differentiation		[140]
GRα	Transcription factor, Nuclear receptor signaling	Ser404	[146]
MED13/13L	Mediator, Transcriptional regulation	Thr326, Ser330	[82]
Mcl-1	Bcl-family protein, Anti-apoptotic signaling	Ser64, Ser121, Ser159	[117, 142]
mTOR	Protein kinase, PI3K signaling	Thr631	[48]
B-Raf	Protein kinase, ERK signaling	Thr401, Ser405	[112]
Aurora A	Protein kinase, Cell cycle	Thr217	[134]
Aurora B	Protein kinase, Cell cycle		[113]
Presenilin	Protease, Notch signaling	Thr116	[122]
SRC-3	Transcriptional co-activator, Circadian rhythm	Ser101, Ser102	[123]
PGC-1α	Transcriptional co-activator	Thr255, Thr295	[127]
DEK	Chromatin regulator, Chromatin modification and mRNA splicing	Thr15, Thr66	[126]
TGIF1	Transcription factor, TGFβ signaling	Thr235	[231]
TopoIIα	Topoisomerase, Mitosis and meiosis	Ser1361	[232]
NF1	Regulator of Ras GTPase activity, Ras signaling pathway	Thr2757	[233]
RCAN1	Calcineurin A binding protein, Calcineurin signaling pathway		[234]
CCDC6	ATM substrate, DNA damage response	Ser359, Ser413, Thr427	[235]
GCSF receptor	Receptor, Cytokine signaling pathway		[236]

FBW7α. Three novel exons, located upstream of the previously reported first exon, could constitute seven additional *FBW7α* splicing isoforms [54].

Structurally, the three protein isoforms contain a dimerization domain (DD), an F-box domain, seven WD-40 repeats, and a unique isoform-specific N-terminus, which may provide signals regulating subcellular localization, protein expression pattern as well as function [1]. The DD domain ranges from amino acids 145 to 193, which is required for FBW7 dimerization. The linked FBW7 dimers orientate suprafacially and accommodate multiple geometries for its substrate recognition [57]. It is believed that the heterodimeric FBW7 complex comprises the binding pocket to which phosphorylated substrates might interact [57]. Similar to other F-box proteins, FBW7 is responsible for recruiting specific substrates to the SCF core complex in a phosphorylation-dependent manner. The C-terminal WD domain contains seven WD-40 repeats that form a seven-bladed, barrel-shaped β-propeller structure to form a phospho-degron binding pocket, by which FBW7 recognizes and binds to conserved phospho-motifs present within substrates [57]. Most FBW7 targets contain a single high affinity phospho-motif (I/L-I/L/P-T/S-X-X-S/T/E), which is termed the Cdc4 phospho-degron (CPD) [1].

As mentioned above, the unique N-terminus provides each isoform with a distinct subcellular distribution as well as expression pattern [58]. Specifically, FBW7α is ubiquitously expressed in human tissues and localizes to the nucleoplasm, whereas FBW7β is a primarily cytoplasmic protein and enriched in the brain and thymus. FBW7γ is restricted to heart and skeletal muscle and shows a nucleolar distribution [41, 49, 55, 48]. Owing to the high and ubiquitous expression pattern of FBW7α [49, 55], it is reasoned that FBW7α is the major isoform targeting substrates for degradation. This has recently been confirmed by Grim et al., who established isoform-specific *FBW7*-null mutations in human cells, and found that FBW7α is responsible for the degradation of most FBW7 substrates [59]. Additionally, in spite of the subcellular distribution differences, each isoform showed functional compensation in degrading substrates [60]. These functional redundancies and localization differences offer another layer of complexity for the analysis of FBW7 functions. Although the Reed group reported the collaboration of FBW7α and FBW7γ in promoting Cyclin E ubiquitination, more evidence is still necessary to understand the cooperation of each isoform in FBW7 physiological function [61, 62]. It is likely that additional determinants mediate the interaction between substrate and a specific isoform.

2.2.2 Oncogenic Substrates of FBW7

Most FBW7 substrates are vital regulatory effectors involved in various cellular processes (Table 2.5). For instance, Cyclin E participates in cell cycle regulation [63]; c-Myc and c-Jun are critical for cell cycle progression and cell size determination [64, 65]; Notch functions as a regulator of cell fate determination and differentiation [66]; and Mcl-1 is an anti-apoptotic Bcl-2 superfamily member [67]. In this section, we will discuss several key substrates that will allow us to understand the critical role of FBW7 in cell cycle control, apoptosis, tumor metastasis, and drug resistance.

2.2.2.1 Cyclin E

Cyclin E is the best-characterized substrate of FBW7. It binds to and activates Cyclin-dependent kinase 2 (Cdk2) and catalyzes the transition from the G1 phase to the S phase of the cell cycle [67]. As a critical regulator of cell cycle procession, the amount of Cyclin E present in the cell is tightly controlled by ubiquitin-mediated proteolysis [68]. FBW7 associates specifically with phosphorylated Cyclin E, and targets Cyclin E for ubiquitination and subsequent degradation by the 26S proteasome [69]. Deregulation of Cyclin E has been frequently found in cancer, and enhanced expression of Cyclin E leads to genomic instability and tumorigenesis [68]. Depletion of FBW7 leads to accumulation and stabilization of Cyclin E in various types of human malignances [38]. Thus, Cyclin E is considered a key mediator of FBW7 tumor suppression. In addition, Cyclin E-induced genomic instability in primary human cells can be prevented by cooperation of the p53 and FBW7 pathways, in which overexpression of Cyclin E activates p53, which in turn inhibits Cdk2/Cyclin E activity by induction of $p21^{Cip1}$ [70]. This finding was further confirmed by Minella et al., who demonstrated that mutant Cyclin E, which cannot be degraded by FBW7, induced genomic instability [69]. More recently, one report has defined Cyclin E as the critical signaling connector by which FBW7 governs APC^{Cdh1} activity [71]. Specifically, depletion of Cyclin E in FBW7-deficient cells reduced the expression of elevated APC^{Cdh1} substrates to levels comparable to those in wild-type cells. Conversely, overexpression of Cyclin E recapitulated aberrant Cdh1 substrate expression, which was originally observed in FBW7-deficient cells [71, 72]. More importantly, a Cdh1 mutant that is resistant to Cdk2/Cyclin E-mediated phosphorylation reversed the elevated expression of various APC^{Cdh1} substrates in FBW7-deficient cells [71]. This finding suggests that Cdk2/Cyclin E inhibitors could serve as effective therapeutic agents for treating FBW7-deficient tumors.

Interestingly, two CPD consensus sequences are found within Cyclin E, which has puzzled researchers for quite some time. One CPD is located in the N-terminus and contains a Thr62 phosphorylation site while another is found in the C-terminus and contains Thr380 and Ser384 phosphorylation sites. T380 is phosphorylated by both the Cdk2 and GSK3 kinases, and the phosphorylated T380 degron directly binds to FBW7 [38]. As such, mutation of T380 disrupts Cyclin E ubiquitylation in vivo and in vitro [73]. Moreover, S384 is uniquely phosphorylated by Cdk2 and this phosphorylation provides a negative charge at the +4 position to further increase the binding affinity to FBW7 [74]. Recent work by the Reed group has begun to shed some light on the importance of these two CPDs by showing that degradation of Cyclin E in vivo requires both FBW7α and FBW7γ [62]. In their model, FBW7α serves as a cofactor for the prolyl *cis–trans* isomerase Pin1 in the isomerization of Cyclin E. Pin1-mediated isomerization of Cyclin E, and subsequent binding to FBW7α, then drives nucleolar localization of Cyclin E, where it is ubiquitylated by FBW7γ prior to its degradation by the proteasome [62]. However, more evidence is still needed to better understand the substrate specificity of each FBW7 isoform, and to further define the role of each CPD in Cyclin E turnover.

2.2.2.2 Notch

Mammals express four transmembrane Notch receptors (Notch-1, -2, -3 and -4) and five canonical transmembrane ligands (DLL 1, DLL 3, DLL 4, Jagged-1, and Jagged-2) [75]. Deregulated expression of Notch proteins, ligands and targets, including overexpression and activation of Notch, has been reported in multiple solid tumors, including cervical [76], lung [77], pancreatic [78], hepatocellular [79], gastric carcinomas [80] and melanoma [81]. Interestingly, Notch is a well-characterized substrate of FBW7. Work from the Nakayama and Elledge groups have described a molecular mechanism by which Notch degradation is mediated by FBW7 in mice [40, 82]. Furthermore, FBW7 was first identified in *C. elegans* as a negative regulator of Notch by genetic screening [83]. Fryer et al. showed that phosphorylation of Notch in the PEST (rich in amino acids P, E, S and T) domain by Cyclin C/Cdk8 leads to binding of FBW7, resulting in turnover of the complex [84]. O'Neil et al. have also defined a functional CPD within Notch at Thr2512 [85]. In support of Notch being a FBW7 substrate, mice lacking *Fbw7* exhibited elevated Notch expression and subsequently impaired cardiovascular development [40]. Moreover, *FBW7* mutations in leukemic cells were found to result in Notch pathway activation through inhibition of Notch degradation [85]. Recently, it has also been demonstrated that SGK1 (serum- and glucocorticoid-inducible protein kinase 1) significantly reduced Notch stability through FBW7 [86]. Moreover, the intracellular domain of Jagged-1 was also found to interact with the Notch1 intra-cellular domain and promote its degradation through a FBW7-dependent protea-somal pathway [87].

2.2.3 Aberrant Regulation of FBW7 in Cancer Progression

Although intensive efforts have been made to identify various downstream ubiquitin targets for FBW7, relatively little is known about the upstream signaling pathways that control FBW7 stability and cellular functions as well as the regulation of FBW7 itself. To this end, there is emerging evidence demonstrating that the tumor suppressor functions of FBW7 could be governed by multiple regulators, mutations, alternative splicing as well as upstream cellular signaling pathways.

2.2.3.1   Regulation of FBW7 by Upstream Genes

Recently, *FBW7* has been identified as a *bona-fide* transcriptional target of p53 by Mao and colleagues [54]. p53 is a tumor suppressor and involved in many cellular processes including cell growth, DNA synthesis and repair, differentiation, apoptosis and cellular response to a wide range of biological stresses. Initially, Kimura et al. found that expression of *FBW7* was dramatically upregulated by infecting p53-deficient cells with an adenovirus encoding wild-type p53 [88]. Furthermore, they demonstrated that the first exon of *FBW7* contains a p53-binding site that displays p53-dependent transcriptional activity [88]. In addition, expression of FBW7β was induced in a p53-dependent manner after genotoxic stresses such as UV irradiation, suggesting that *FBW7* is a direct transcriptional target of p53. In line with this finding, Mao et al. subsequently reported that FBW7 mediated the role of p53 in response to DNA damage, indicating that the *FBW7* gene is a p53-dependent tumor suppressor. Taken together, these studies showed that targeting the p53 signaling pathway could potentially influence FBW7 expression, which might provide a feasible approach to restore FBW7 expression during anticancer therapies [54].

Additionally, Pawar et al. reported that the transcription factor C/EBPδ (CCAAT/enhancer binding protein δ) directly inhibited *FBW7* gene expression and induced the accumulation of FBW7 oncogenic targets mTOR and Aurora A. C/EBPδ is one of six isoforms of the C/EBP family that is a highly conserved family of leucine zipper type DNA-binding proteins [89]. By binding to the transcription motif of the *FBW7* promoter, C/EBPδ may block the access of p53, resulting in suppression of *FBW7* transcription [90]. Interestingly, this study also revealed that C/EBPδ promoted breast tumor metastasis, indicating that further investigation is required to determine the molecular mechanisms, especially the contribution of FBW7 in mediating the cellular functions of C/EBPδ in promoting tumor metastasis [91].

Another positive regulator of FBW7, Numb, has also been recently identified. Numb is a membrane-bound protein that associates with Notch-1. In several types of cancers, loss of Numb expression has been observed [92–94]. In approximately half of all human mammary carcinomas, Numb-mediated suppression of Notch signaling is lost in part due to Numb ubiquitination and proteasomal degradation [95]. Recently, it has been shown that one of the predominant Numb isoforms, Numb4, promoted SCF^FBW7 ubiquitin ligase assembly and activation, leading to enhanced Notch degradation [96]. However, further in-depth investigation is required to understand the physiological contribution of Numb4-mediated regulation of SCF^FBW7 ubiquitin ligase activity in tumorigenesis in vivo.

2.2.3.2   Regulation of FBW7 by the MicroRNAs (miRNAs)

Another potential group of *FBW7* upstream regulators are micro-RNAs. microRNAs are short RNA molecules that average 22 nucleotides long and bind to complementary sequences on target mRNAs, resulting in translational repression or target degradation, leading to gene silencing [97–99]. Recent studies have shown that multiple miRNAs including miR-27a and miR-223 could regulate FBW7 expression.

Xu et al. first reported that miR-223 could directly regulate the activity of the SCF[FBW7] ubiquitin ligase [100]. They found that overexpression of miR-223 could increase endogenous Cyclin E protein abundance and activity, leading to increased genomic instability by significantly reducing *FBW7* mRNA levels. Conversely, reduced miR-223 expression resulted in increased FBW7 expression and decreased Cyclin E activity, indicating that *FBW7* could be modulated directly by miR-223 [100, 101]. Consistent with this finding, negative regulation of *FBW7* by miR-223 was also described in gastric tumor tissues, squamous cell carcinoma and esophageal cancers [53, 102]. These data indicate that microRNAs may be involved in the transcriptional regulation of *FBW7* to affect FBW7 substrate degradation and function.

It has also been documented that miR-27a played an oncogenic role in human cancers [103–105]. An inverse correlation between miR-27a expression and FBW7 levels in human tumor samples has been observed, indicating that *FBW7* is a potential target of miR-27a [104, 106]. Recently, Lerner et al. identified that miR-27a suppressed FBW7 expression, leading to a reduction in ubiquitin-mediated degradation and turnover of the FBW7 substrate, Cyclin E [107]. Overexpression of FBW7 caused dysregulation of Cyclin E, resulting in altered cell cycle progression. Conversely, miR-27a knockdown increased FBW7 levels and subsequently decreased the abundance of FBW7 substrates such as c-Myc, c-Jun and Notch-1 in colon cancer derived cell lines. Furthermore, miR-27a overexpression promoted cell growth, whereas miR-27a knockdown inhibited cell proliferation in vitro and tumor formation in vivo in part through regulating FBW7. In addition, miR-27a was found to primarily suppressed FBW7 expression during G1–S phase transition [107].

In addition, miR-25 has also been shown to inhibit FBW7 expression and caused upregulation of c-Myc and KLF5 to promote reprogramming of mouse fibroblast cells to iPSCs [108]. Overexpression of miR-129-5p led to upregulation of FBW7 expression [109]. However, it remains largely unclear how these different miRNAs regulate *FBW7* in a redundant or synergistic manner, and whether different subsets of miRNAs play a primary role in different tissue or cellular context.

2.2.3.3 Regulation of Translational Efficiency by Alternative Splicing

Largely functioning as a tumor suppressor to promote the degradation of multiple oncoproteins, loss of FBW7 expression is frequently observed in various human cancers [50]. However, the exact molecular mechanisms regulating FBW7 expression still remain poorly understood. Using the RACE technique, Liu et al. examined the 5′ region of the *FBW7* gene and identified three novel noncoding exons located in the 5′ untranslated region (UTR). Notably, this region is composed of seven alternatively spliced 5′-UTR forms of *FBW7α* and significant differences in the translational efficiency among these 5′-UTR variants were observed. Furthermore, the mRNA levels of these splice forms were reduced in more than 80 % of all breast cancer cell lines tested and in more than 50 % of all human primary cancers examined from various tissues. These results suggested that differential expression of *FBW7α* splice forms with different translational properties may serve as a novel mechanism for inactivation of FBW7 in human cancers [49].

2.2.3.4 Dominant Negative Regulation of FBW7 Mutations

FBW7 status has been examined among numerous primary human tumors [110]. In a comprehensive study of over 1,500 human tumors, Akhoondi et al. found that 6 % of tumors examined harbored mutations in the *FBW7* coding regions. Cholangiocarcinoma and T-ALL contained the highest frequency of mutations at 35 % and 31 %, respectively, while only 9–15 % of tumors of the stomach, colon, pancreas, and endometrium contained detectable *FBW7* mutations [54]. Strikingly, nearly half (43 %) of these were missense mutations and resulted in amino acid substitutions at key arginine residues within the WD40 domain (Arg465, Arg479, and Arg505), which are shared by all three FBW7 isoforms. Most of the remaining mutations were nonsense codons that lead to premature termination of FBW7 translation.

Unlike many known tumor suppressors, FBW7 tends to be mutated in only one allele [54]. The mutations including both missense and nonsense mutations result in the inactivation of FBW7 by loss of binding capability to substrates. Impaired substrate binding in full-length FBW7 mutants raises the possibility that these mutations might function as potent dominant negatives. Indeed, several groups have shown that FBW7 hot-spot mutants can dominantly interfere with the ability of the wild-type protein to degrade c-Myc, Sic1, and Cyclin E [50, 57, 85]. Dominant negative FBW7 mutants form heterodimers with wild-type FBW7, and therefore causing loss of binding between dimerized SCF^FBW7 E3 ligases and target substrates, making FBW7 a haplo-insufficient tumor suppressor [1, 111].

2.2.4 FBW7 Knockout Mouse Models: Implications for Tumor Development

To better understand the underlying mechanisms of tumor formation due to loss of FBW7 function, several FBW7 conditional knockout and knock-in mouse models have been developed (Tables 2.1 and 2.3).

Double allele knockout of *FBW7* is lethal and causes embryos to die in utero at embryonic day 10.5 with growth retardation and impaired vascular development, which might be in part caused by the accumulation of FBW7 substrates Notch-1 and Notch-4 [40, 41]. Due to the limitation of embryonic mortality associated with the *Fbw7*^−/− mice, conditional ablation of *Fbw7* in various adult tissues has been developed. So far, conditional inactivation of *Fbw7* in the T-cell lineage, bone marrow, intestine, liver, breast, and brain has been reported (Table 2.1).

Specifically, mice with conditional inactivation of *Fbw7* in the T-cell lineage develop thymic lymphoma partly due to excessive accumulation of c-Myc [112]. Bone marrow (BM)-specific *Fbw7* knockout mice exhibit extremely severe pancytopenia 12 weeks post-deletion of *Fbw7*, and develop T-ALL within 16 weeks due to the accumulation of Notch and c-Myc as well as deregulation of p53-induced exhaustion of hematopoietic stem cells (HSCs) [82, 113]. Interestingly, King et al. demonstrated that, in contrast to these knockout mouse models, HSCs with a *Fbw7* WD40 mutation (*Fbw7$^{R465Cl+}$*) did not compromise HSC functions but increased the proportion of leukemia initiating cells in collaboration with Notch1 in vivo [114]. Mice with *Fbw7* deletion in the gut survived and were fertile without exhibiting any gross phenotypic alteration [115]. However, when crossed with *APC$^{min/+}$* mice, intestinal tumors developed due to enhanced APC (Adenomatous polyposis coli-mediated tumorigenesis)-mediated intestinal tumorigenesis via upregulation of the FBW7 substrates Notch-1 and c-Jun [48]. More recent studies have also demonstrated that mice harboring a heterozygous hot-spot mutant (R482Q, which is equivalent of R479Q in human) in the *APC* mutant background (*Fbw7$^{R482Q/+}$/Apc$^{1322T/+}$*) developed intestinal tumor more frequently than *Fbw7* heterozygous mice in the same *Apc$^{1322T/+}$* background [113, 146]. Furthermore, *Fbw7* deficiency in the liver eventually caused the development of hematomas partly due to accumulation of SREBP and Notch-1 [121]. In addition, *Fbw7$^{+/-}$* mice are susceptible to radiation-induced tumorigenesis and irradiated *Fbw7$^{+/-}$/p53$^{+/-}$* mice developed lymphomas in 70 % of total mice examined [54]. These mice also displayed a wide spectrum of different tissue tumors in lung, liver, and ovary [54]. Taken together, FBW7 functions as a haploid insufficient tumor suppressor with single allele deletion capable of promoting tumorigenesis.

2.2.5   Targeting the FBW7 Signaling Pathway as a Potential Therapeutic Anticancer Strategy

Although FBW7 behaves as a tumor suppressor, emerging evidence points to FBW7 as a potential therapeutic target. First, FBW7 functions as a pro-survival factor in multiple myeloma by constitutively targeting the NF-κB inhibitor p100 for degradation in a GSK3-dependent manner [118–120]. NF-κB plays a key role in regulating the immune response to infections and is involved in cellular responses to stimuli such as stress, cytokines, and free radicals. In addition, aberrant regulation of NF-κB has been linked to cancer [121]. Second, a number of cancers, including T-ALL, breast cancers and gastric adenocarcinoma often carry mutations in the *FBW7* gene, leading to an accumulation of mitogenic proteins, whereas these mutations do not occur in B-cell malignancies like multiple myeloma [50]. Therefore, FBW7 and GSK3 may serve as promising targets for the treatment of multiple myelomas with constitutive activation of the NF-κB pathway.

More recent reports have revealed that FBW7 plays a pivotal role in the maintenance of quiescence in leukemia-initiating cells (LICs) of chronic myeloid leukemia (CML) [114]. Ablation of *Fbw7* in LICs leads to accumulation of c-Myc and impaired maintenance of quiescence followed by p53-dependent apoptosis and cellular exhaustion. Furthermore, small-molecule mediated suppression of Myc activity results in the T-ALL remission. *Fbw7* deletion induces LICs into cell cycle progress and thus sensitizes the LICs to Imatinib (Gleevec) treatment [122, 123]. These studies identify FBW7 as an essential regulator of CML's LIC maintenance and open the way for targeting FBW7 activity in CML.

Furthermore, given the frequent loss of FBW7 in various human cancers and the fact that many FBW7 downstream targets are oncoproteins, it will be important to identify the driving oncoproteins in different types of FBW7-deficient human cancers. As these FBW7-deficient tumors are addicted to these driving oncoproteins, targeting specific oncoproteins regulated by FBW7 will be more efficient for cancer treatment.

2.3 The Emerging Tumor Suppressor Role of FBXW8

FBXW8 (also known as FBW6, FBW8, FBX29, FBXW6, or FBXO29) is the only substrate receptor identified for the Cul7-SCF E3 ubiquitin ligase complex. FBXW8 was shown to play a pivotal role in cancer cell proliferation in part by promoting Cyclin D1 degradation during S phase [124], and the turnover of IRS1, a key component of signaling pathways activated by the insulin and insulin-like growth factor 1 (IGF-1) receptors (Table 2.6). In this regard, a recent study showed that mTORC2

Table 2.6 Summary of identified ubiquitin substrates for the FBXW class of F-box proteins other than β-TRCP1/2 and Fbxw7

F-box protein	Substrate	Signaling pathway/functions	References
FBXW2 Undetermined	GCMa	Transcription factor critical for glial and placental cell differentiation	[237]
FBXW5 Undetermined	Eps8	A bifunctional actin cytoskeleton remodeller, a positive regulator of cell proliferation and motility	[238]
	HsSAS-6	Centriolar protein	[239, 240]
FBXW8 Emerging tumor suppressor	IRS-1	An adaptor protein that is one of the major substrates of the insulin receptor kinase	[125]
	TBC1D3	A GTPase activating protein for RAB5	[241]
	Cyclin D1	Cell cycle	[124]
	GRASP65	Stacking factor involved in the postmitotic assembly of Golgi stacks from mitotic Golgi fragments	[129]
	IGFBP2	IGF-binding protein	[127]
FBXW10 Undetermined	HP1alpha and beta	Heterochromatin protein 1alpha and beta	[242]
FBXW15 Undetermined	HBO1	Origin recognition complex, DNA replication licensing	[243]

stabilizes FBXW8 by phosphorylation of Ser86, allowing the insulin-induced translocation of FBXW8 to the cytosol where it mediates IRS-1 degradation [125]. In spite of these studies, it remains unclear whether FBXW8 is necessary for cell cycle progression in normal cells.

FBXW8 is not expressed at a detectable level in adult mouse tissues with the exception of the placenta [126, 127]. Moreover, FBXW8 is dispensable for Cyclin D1 degradation in MEFs [128]. Together, these observations suggest that FBXW8 might be required for Cyclin D1 degradation only in certain cancer cell lines. Two independent groups analyzed the function of FBXW8 in vivo by generating *Fbxw8* knockout mice (Table 2.1). In the first study, exon 2 that encodes the F-box domain was disrupted, leading to approximately 60 % of *Fbxw8$^{-/-}$* embryos dying in utero, where remaining embryos are born alive and grow to the adulthood [126]. In the second study, a gene-trap approach was used with the targeting vector inserted at the 3′ end of exon 3 of the *Fbxw8* gene. Again, about 30 % of *Fbxw8$^{-/-}$* mice survived after the birth, but these mice remained smaller in both body weight and organ sizes compared with their wild-type littermates [127]. Phenotypically, both studies showed that *Fbxw8*-deficient mice exhibited prenatal and postnatal growth retardation resulting from a defect in placental development [126, 127]. Given that FBXW8 is dispensable for the degradation of Cyclin D1 in MEFs, the phenotype of Fbxw8-deficient mice may be due to the accumulation of substrates other than Cyclin D1 (Table 2.1). These studies suggest that the accumulation of IRS-1 or other still unknown FBXW8 substrates in the placenta might be responsible for the growth retardation observed in mice deficient in *Fbxw8*.

Additionally, the Cul7^{FBXW8} complex was found to control the morphogenesis of the Golgi apparatus and patterning of dendrites by targeting for ubiquitination the Golgi protein Grasp65 [129]. These findings link the Cul7^{FBXW8} ubiquitin signaling mechanism with the normal development of the brain [129]. Together, these studies revealed FBXW8 as a key factor controlling organismal growth and development. However, more thorough studies are required to further understand the physiological role of FBXW8 in tumorigenesis by generating additional conditional knockout or transgenic mouse models as well as biochemically identifying additional ubiquitin substrates for Cul7^{FBXW8} to reveal the major signaling pathways controlled by Cul7^{FBXW8}.

References

1. Welcker M, Clurman BE. FBW7 ubiquitin ligase: a tumour suppressor at the crossroads of cell division, growth and differentiation. Nat Rev Cancer. 2008;8(2):83–93.
2. Fuchs SY, Spiegelman VS, Kumar KG. The many faces of beta-TrCP E3 ubiquitin ligases: reflections in the magic mirror of cancer. Oncogene. 2004;23(11):2028–36.
3. Margottin F, et al. A novel human WD protein, h-beta TrCp, that interacts with HIV-1 Vpu connects CD4 to the ER degradation pathway through an F-box motif. Mol Cell. 1998;1(4): 565–74.
4. Suzuki H, et al. Homodimer of two F-box proteins betaTrCP1 or betaTrCP2 binds to IkappaBalpha for signal-dependent ubiquitination. J Biol Chem. 2000;275(4):2877–84.

5. Bai C, et al. SKP1 connects cell cycle regulators to the ubiquitin proteolysis machinery through a novel motif, the F-box. Cell. 1996;86(2):263–74.
6. Hart M, et al. The F-box protein beta-TrCP associates with phosphorylated beta-catenin and regulates its activity in the cell. Curr Biol. 1999;9(4):207–10.
7. Yaron A, et al. Identification of the receptor component of the IkappaBalpha-ubiquitin ligase. Nature. 1998;396(6711):590–4.
8. Nakayama K, et al. Impaired degradation of inhibitory subunit of NF-kappa B (I kappa B) and beta-catenin as a result of targeted disruption of the beta-TrCP1 gene. Proc Natl Acad Sci U S A. 2003;100(15):8752–7.
9. Margottin-Goguet F, et al. Prophase destruction of Emi1 by the SCF(betaTrCP/Slimb) ubiquitin ligase activates the anaphase promoting complex to allow progression beyond prometaphase. Dev Cell. 2003;4(6):813–26.
10. Busino L, et al. Degradation of Cdc25A by beta-TrCP during S phase and in response to DNA damage. Nature. 2003;426(6962):87–91.
11. Jin J, et al. SCFbeta-TRCP links Chk1 signaling to degradation of the Cdc25A protein phosphatase. Genes Dev. 2003;17(24):3062–74.
12. Watanabe N, et al. M-phase kinases induce phospho-dependent ubiquitination of somatic Wee1 by SCFbeta-TrCP. Proc Natl Acad Sci U S A. 2004;101(13):4419–24.
13. Wei S, et al. A novel mechanism by which thiazolidinediones facilitate the proteasomal degradation of cyclin D1 in cancer cells. J Biol Chem. 2008;283(39):26759–70.
14. Sasajima H, et al. Polyubiquitination of the B-cell translocation gene 1 and 2 proteins is promoted by the SCF ubiquitin ligase complex containing betaTrCP. Biol Pharm Bull. 2012;35(9):1539–45.
15. Wojcik EJ, Glover DM, Hays TS. The SCF ubiquitin ligase protein slimb regulates centrosome duplication in Drosophila. Curr Biol. 2000;10(18):1131–4.
16. Zhou BP, et al. Dual regulation of Snail by GSK-3beta-mediated phosphorylation in control of epithelial-mesenchymal transition. Nat Cell Biol. 2004;6(10):931–40.
17. Ray D, Osmundson EC, Kiyokawa H. Constitutive and UV-induced fibronectin degradation is a ubiquitination-dependent process controlled by beta-TrCP. J Biol Chem. 2006;281(32):23060–5.
18. Ding Q, et al. Degradation of Mcl-1 by beta-TrCP mediates glycogen synthase kinase 3-induced tumor suppression and chemosensitization. Mol Cell Biol. 2007;27(11):4006–17.
19. Dehan E, et al. betaTrCP- and Rsk1/2-mediated degradation of BimEL inhibits apoptosis. Mol Cell. 2009;33(1):109–16.
20. Dorrello NV, et al. S6K1- and betaTRCP-mediated degradation of PDCD4 promotes protein translation and cell growth. Science. 2006;314(5798):467–71.
21. Tan M, et al. SAG/ROC-SCF beta-TrCP E3 ubiquitin ligase promotes pro-caspase-3 degradation as a mechanism of apoptosis protection. Neoplasia. 2006;8(12):1042–54.
22. Ougolkov A, et al. Associations among beta-TrCP, an E3 ubiquitin ligase receptor, beta-catenin, and NF-kappaB in colorectal cancer. J Natl Cancer Inst. 2004;96(15):1161–70.
23. Koch A, et al. Elevated expression of Wnt antagonists is a common event in hepatoblastomas. Clin Cancer Res. 2005;11(12):4295–304.
24. Muerkoster S, et al. Increased expression of the E3-ubiquitin ligase receptor subunit betaTRCP1 relates to constitutive nuclear factor-kappaB activation and chemoresistance in pancreatic carcinoma cells. Cancer Res. 2005;65(4):1316–24.
25. Liu H, Cheng EH, Hsieh JJ. Bimodal degradation of MLL by SCFSkp2 and APCCdc20 assures cell cycle execution: a critical regulatory circuit lost in leukemogenic MLL fusions. Genes Dev. 2007;21(19):2385–98.
26. Kudo Y, et al. Role of F-box protein betaTrcp1 in mammary gland development and tumorigenesis. Mol Cell Biol. 2004;24(18):8184–94.
27. Belaidouni N, et al. Overexpression of human beta TrCP1 deleted of its F box induces tumorigenesis in transgenic mice. Oncogene. 2005;24(13):2271–6.
28. Tsai WB, et al. Inhibition of FOXO3 tumor suppressor function by betaTrCP1 through ubiquitin-mediated degradation in a tumor mouse model. PLoS One. 2010;5(7):e11171.

29. Saitoh T, Katoh M. Expression profiles of betaTRCP1 and betaTRCP2, and mutation analysis of betaTRCP2 in gastric cancer. Int J Oncol. 2001;18(5):959–64.
30. Kim CJ, et al. Somatic mutations of the beta-TrCP gene in gastric cancer. APMIS. 2007;115(2):127–33.
31. Gerstein AV, et al. APC/CTNNB1 (beta-catenin) pathway alterations in human prostate cancers. Genes Chromosomes Cancer. 2002;34(1):9–16.
32. Wood LD, et al. The genomic landscapes of human breast and colorectal cancers. Science. 2007;318(5853):1108–13.
33. Shirane M, et al. Common pathway for the ubiquitination of IkappaBalpha, IkappaBbeta, and IkappaBepsilon mediated by the F-box protein FWD1. J Biol Chem. 1999;274(40): 28169–74.
34. Tan P, et al. Recruitment of a ROC1-CUL1 ubiquitin ligase by Skp1 and HOS to catalyze the ubiquitination of I kappa B alpha. Mol Cell. 1999;3(4):527–33.
35. Winston JT, et al. The SCFbeta-TRCP-ubiquitin ligase complex associates specifically with phosphorylated destruction motifs in IkappaBalpha and beta-catenin and stimulates IkappaBalpha ubiquitination in vitro. Genes Dev. 1999;13(3):270–83.
36. Xia Y, et al. Phosphorylation of p53 by IkappaB kinase 2 promotes its degradation by beta-TrCP. Proc Natl Acad Sci U S A. 2009;106(8):2629–34.
37. Simchen G, Hirschberg J. Effects of the mitotic cell-cycle mutation cdc4 on yeast meiosis. Genetics. 1977;86(1):57–72.
38. Koepp DM, et al. Phosphorylation-dependent ubiquitination of cyclin E by the SCFFbw7 ubiquitin ligase. Science. 2001;294(5540):173–7.
39. Yada M, et al. Phosphorylation-dependent degradation of c-Myc is mediated by the F-box protein Fbw7. EMBO J. 2004;23(10):2116–25.
40. Tetzlaff MT, et al. Defective cardiovascular development and elevated cyclin E and Notch proteins in mice lacking the Fbw7 F-box protein. Proc Natl Acad Sci U S A. 2004; 101(10):3338–45.
41. Tsunematsu R, et al. Mouse Fbw7/Sel-10/Cdc4 is required for notch degradation during vascular development. J Biol Chem. 2004;279(10):9417–23.
42. Nateri AS, et al. The ubiquitin ligase SCFFbw7 antagonizes apoptotic JNK signaling. Science. 2004;303(5662):1374–8.
43. Wei W, et al. The v-Jun point mutation allows c-Jun to escape GSK3-dependent recognition and destruction by the Fbw7 ubiquitin ligase. Cancer Cell. 2005;8(1):25–33.
44. Inuzuka H, et al. SCF(FBW7) regulates cellular apoptosis by targeting MCL1 for ubiquitylation and destruction. Nature. 2011;471(7336):104–9.
45. Wertz IE, et al. Sensitivity to antitubulin chemotherapeutics is regulated by MCL1 and FBW7. Nature. 2011;471(7336):110–4.
46. Mao JH, et al. FBXW7 targets mTOR for degradation and cooperates with PTEN in tumor suppression. Science. 2008;321(5895):1499–502.
47. Thompson BJ, et al. The SCFFBW7 ubiquitin ligase complex as a tumor suppressor in T cell leukemia. J Exp Med. 2007;204(8):1825–35.
48. Sancho R, et al. F-box and WD repeat domain-containing 7 regulates intestinal cell lineage commitment and is a haploinsufficient tumor suppressor. Gastroenterology. 2010;139(3): 929–41.
49. Liu B, et al. Proteomic identification of common SCF ubiquitin ligase FBXO6-interacting glycoproteins in three kinds of cells. J Proteome Res. 2012;11(3):1773–81.
50. Akhoondi S, et al. FBXW7/hCDC4 is a general tumor suppressor in human cancer. Cancer Res. 2007;67(19):9006–12.
51. Koh MS, et al. CDC4 gene expression as potential biomarker for targeted therapy in prostate cancer. Cancer Biol Ther. 2006;5(1):78–83.
52. Cheng Y, Li G. Role of the ubiquitin ligase Fbw7 in cancer progression. Cancer Metastasis Rev. 2012;31(1–2):75–87.
53. Li J, et al. MicroRNA-223 functions as an oncogene in human gastric cancer by targeting FBXW7/hCdc4. J Cancer Res Clin Oncol. 2012;138(5):763–74.

54. Mao JH, et al. Fbxw7/Cdc4 is a p53-dependent, haploinsufficient tumour suppressor gene. Nature. 2004;432(7018):775–9.
55. Welcker M, et al. The Fbw7 tumor suppressor regulates glycogen synthase kinase 3 phosphorylation-dependent c-Myc protein degradation. Proc Natl Acad Sci U S A. 2004; 101(24):9085–90.
56. Liu Y, et al. Multiple novel alternative splicing forms of FBXW7alpha have a translational modulatory function and show specific alteration in human cancer. PLoS One. 2012;7(11):e49453.
57. Tang X, et al. Suprafacial orientation of the SCFCdc4 dimer accommodates multiple geometries for substrate ubiquitination. Cell. 2007;129(6):1165–76.
58. Spruck CH, et al. hCDC4 gene mutations in endometrial cancer. Cancer Res. 2002;62(16): 4535–9.
59. Grim JE, et al. Fbw7 and p53 cooperatively suppress advanced and chromosomally unstable intestinal cancer. Mol Cell Biol. 2012;32(11):2160–7.
60. Reed SI. Cooperation between different Cdc4/Fbw7 isoforms may be associated with 2-step inactivation of SCF(Cdc4) targets. Cell Cycle. 2006;5(17):1923–4.
61. van Drogen F, et al. Ubiquitylation of cyclin E requires the sequential function of SCF complexes containing distinct hCdc4 isoforms. Mol Cell. 2006;23(1):37–48.
62. Bhaskaran N, et al. Fbw7alpha and Fbw7gamma collaborate to shuttle cyclin E1 into the nucleolus for multiubiquitylation. Mol Cell Biol. 2013;33(1):85–97.
63. Siu KT, Rosner MR, Minella AC. An integrated view of cyclin E function and regulation. Cell Cycle. 2012;11(1):57–64.
64. Gallant P. Myc, cell competition, and compensatory proliferation. Cancer Res. 2005;65(15): 6485–7.
65. Fang JY, Richardson BC. The MAPK signalling pathways and colorectal cancer. Lancet Oncol. 2005;6(5):322–7.
66. Espinoza I, et al. Notch signaling: targeting cancer stem cells and epithelial-to-mesenchymal transition. Onco Targets Ther. 2013;6:1249–59.
67. Perciavalle RM, Opferman JT. Delving deeper: MCL-1's contributions to normal and cancer biology. Trends Cell Biol. 2013;23(1):22–9.
68. Spruck CH, Won KA, Reed SI. Deregulated cyclin E induces chromosome instability. Nature. 1999;401(6750):297–300.
69. Minella AC, et al. p53 and SCFFbw7 cooperatively restrain cyclin E-associated genome instability. Oncogene. 2007;26(48):6948–53.
70. Finkin S, et al. Fbw7 regulates the activity of endoreduplication mediators and the p53 pathway to prevent drug-induced polyploidy. Oncogene. 2008;27(32):4411–21.
71. Lau AW, et al. Regulation of APC(Cdh1) E3 ligase activity by the Fbw7/cyclin E signaling axis contributes to the tumor suppressor function of Fbw7. Cell Res. 2013;23(7):947–61.
72. Keck JM, et al. Cyclin E overexpression impairs progression through mitosis by inhibiting APC(Cdh1). J Cell Biol. 2007;178(3):371–85.
73. Strohmaier H, et al. Human F-box protein hCdc4 targets cyclin E for proteolysis and is mutated in a breast cancer cell line. Nature. 2001;413(6853):316–22.
74. Hao B, et al. Structure of a Fbw7-Skp1-cyclin E complex: multisite-phosphorylated substrate recognition by SCF ubiquitin ligases. Mol Cell. 2007;26(1):131–43.
75. Monsalve EM, et al. Abnormal expression pattern of notch receptors, ligands, and downstream effectors in the dorsolateral prefrontal cortex and amygdala of suicidal victims. Mol Neurobiol. Online publication 23 Oct 2013.
76. Ogawa R, et al. NOTCH1 expression predicts patient prognosis in esophageal squamous cell cancer. Eur Surg Res. 2013;51(3–4):101–7.
77. Yi F, Amarasinghe B, Dang TP. Manic fringe inhibits tumor growth by suppressing Notch3 degradation in lung cancer. Am J Cancer Res. 2013;3(5):490–9.
78. Court H, et al. Isoprenylcysteine carboxylmethyltransferase deficiency exacerbates KRAS-driven pancreatic neoplasia via Notch suppression. J Clin Invest. 2013;123(11):4681–94.

79. Morell CM, et al. Notch signalling beyond liver development: emerging concepts in liver repair and oncogenesis. Clin Res Hepatol Gastroenterol. 2013;37(5):447–54.

80. Kang H, et al. Notch3 and Jagged2 contribute to gastric cancer development and to glandular differentiation associated with MUC2 and MUC5AC expression. Histopathology. 2012;61(4): 576–86.

81. Connolly K, et al. Papillomavirus-associated squamous skin cancers following transplant immunosuppression: one Notch closer to control. Cancer Treat Rev. 2013;40(2):205–14.

82. Matsumoto A, et al. Fbxw7-dependent degradation of Notch is required for control of "stemness" and neuronal-glial differentiation in neural stem cells. J Biol Chem. 2011;286(15): 13754–64.

83. Oberg C, et al. The Notch intracellular domain is ubiquitinated and negatively regulated by the mammalian Sel-10 homolog. J Biol Chem. 2001;276(38):35847–53.

84. Fryer CJ, White JB, Jones KA. Mastermind recruits CycC:CDK8 to phosphorylate the Notch ICD and coordinate activation with turnover. Mol Cell. 2004;16(4):509–20.

85. O'Neil J, et al. FBW7 mutations in leukemic cells mediate NOTCH pathway activation and resistance to gamma-secretase inhibitors. J Exp Med. 2007;204(8):1813–24.

86. Mo JS, et al. Serum- and glucocorticoid-inducible kinase 1 (SGK1) controls Notch1 signaling by downregulation of protein stability through Fbw7 ubiquitin ligase. J Cell Sci. 2011;124(Pt 1):100–12.

87. Kim MY, et al. The intracellular domain of Jagged-1 interacts with Notch1 intracellular domain and promotes its degradation through Fbw7 E3 ligase. Exp Cell Res. 2011;317(17): 2438–46.

88. Kimura T, et al. hCDC4b, a regulator of cyclin E, as a direct transcriptional target of p53. Cancer Sci. 2003;94(5):431–6.

89. Tsukada J, et al. The CCAAT/enhancer (C/EBP) family of basic-leucine zipper (bZIP) transcription factors is a multifaceted highly-regulated system for gene regulation. Cytokine. 2011;54(1):6–19.

90. Balamurugan K, et al. FBXW7alpha attenuates inflammatory signalling by downregulating C/EBPdelta and its target gene Tlr4. Nat Commun. 2013;4:1662.

91. Pawar SA, et al. C/EBP{delta} targets cyclin D1 for proteasome-mediated degradation via induction of CDC27/APC3 expression. Proc Natl Acad Sci U S A. 2010;107(20):9210–5.

92. Pece S, et al. NUMB-ing down cancer by more than just a NOTCH. Biochim Biophys Acta. 2011;1815(1):26–43.

93. Colaluca IN, et al. NUMB controls p53 tumour suppressor activity. Nature. 2008;451(7174): 76–80.

94. Misquitta-Ali CM, et al. Global profiling and molecular characterization of alternative splicing events misregulated in lung cancer. Mol Cell Biol. 2011;31(1):138–50.

95. Pece S, et al. Loss of negative regulation by Numb over Notch is relevant to human breast carcinogenesis. J Cell Biol. 2004;167(2):215–21.

96. Jiang X, et al. Numb regulates glioma stem cell fate and growth by altering epidermal growth factor receptor and Skp1-Cullin-F-box ubiquitin ligase activity. Stem Cells. 2012;30(7): 1313–26.

97. Kasinski AL, Slack FJ. Epigenetics and genetics. MicroRNAs en route to the clinic: progress in validating and targeting microRNAs for cancer therapy. Nat Rev Cancer. 2011;11(12): 849–64.

98. Kim M, Kasinski AL, Slack FJ. MicroRNA therapeutics in preclinical cancer models. Lancet Oncol. 2011;12(4):319–21.

99. Kim SE, et al. H-Ras is degraded by Wnt/beta-catenin signaling via beta-TrCP-mediated polyubiquitylation. J Cell Sci. 2009;122(Pt 6):842–8.

100. Xu Y, et al. MicroRNA-223 regulates cyclin E activity by modulating expression of F-box and WD-40 domain protein 7. J Biol Chem. 2010;285(45):34439–46.

101. Mansour MR, et al. The TAL1 complex targets the FBXW7 tumor suppressor by activating miR-223 in human T cell acute lymphoblastic leukemia. J Exp Med. 2013;210(8):1545–57.

102. Kurashige J, et al. Overexpression of microRNA-223 regulates the ubiquitin ligase FBXW7 in oesophageal squamous cell carcinoma. Br J Cancer. 2012;106(1):182–8.
103. Mertens-Talcott SU, et al. The oncogenic microRNA-27a targets genes that regulate specificity protein transcription factors and the G2-M checkpoint in MDA-MB-231 breast cancer cells. Cancer Res. 2007;67(22):11001–11.
104. Wang X, et al. Aberrant expression of oncogenic and tumor-suppressive microRNAs in cervical cancer is required for cancer cell growth. PLoS One. 2008;3(7):e2557.
105. Acunzo M, et al. Cross-talk between MET and EGFR in non-small cell lung cancer involves miR-27a and Sprouty2. Proc Natl Acad Sci U S A. 2013;110(21):8573–8.
106. Spruck C. miR-27a regulation of SCF(Fbw7) in cell division control and cancer. Cell Cycle. 2011;10(19):3232–3.
107. Lerner M, et al. MiRNA-27a controls FBW7/hCDC4-dependent cyclin E degradation and cell cycle progression. Cell Cycle. 2011;10(13):2172–83.
108. Lu D, et al. MiR-25 regulates Wwp2 and Fbxw7 and promotes reprogramming of mouse fibroblast cells to iPSCs. PLoS One. 2012;7(8):e40938.
109. Hasler R, et al. Microbial pattern recognition causes distinct functional micro-RNA signatures in primary human monocytes. PLoS One. 2012;7(2):e31151.
110. Tan Y, Sangfelt O, Spruck C. The Fbxw7/hCdc4 tumor suppressor in human cancer. Cancer Lett. 2008;271(1):1–12.
111. Welcker M, Clurman BE. Fbw7/hCDC4 dimerization regulates its substrate interactions. Cell Div. 2007;2:7.
112. Onoyama I, et al. Conditional inactivation of Fbxw7 impairs cell-cycle exit during T cell differentiation and results in lymphomatogenesis. J Exp Med. 2007;204(12):2875–88.
113. Thompson BJ, et al. Control of hematopoietic stem cell quiescence by the E3 ubiquitin ligase Fbw7. J Exp Med. 2008;205(6):1395–408.
114. King B, et al. The ubiquitin ligase FBXW7 modulates leukemia-initiating cell activity by regulating MYC stability. Cell. 2013;153(7):1552–66.
115. Babaei-Jadidi R, et al. FBXW7 influences murine intestinal homeostasis and cancer, targeting Notch, Jun, and DEK for degradation. J Exp Med. 2011;208(2):295–312.
116. Davis H, et al. Investigation of the atypical FBXW7 mutation spectrum in human tumours by conditional expression of a heterozygous propellor tip missense allele in the mouse intestines. Gut. Online publication 15 May 2013.
117. Onoyama I, et al. Fbxw7 regulates lipid metabolism and cell fate decisions in the mouse liver. J Clin Invest. 2011;121(1):342–54.
118. Busino L, et al. Fbxw7alpha- and GSK3-mediated degradation of p100 is a pro-survival mechanism in multiple myeloma. Nat Cell Biol. 2012;14(4):375–85.
119. Arabi A, et al. Proteomic screen reveals Fbw7 as a modulator of the NF-kappaB pathway. Nat Commun. 2012;3:976.
120. Fukushima H, et al. SCF(Fbw7) modulates the NFkB signaling pathway by targeting NFkB2 for ubiquitination and destruction. Cell Rep. 2012;1(5):434–43.
121. Bellezza I, Mierla AL, Minelli A. Nrf2 and NF-kappaB and their concerted modulation in cancer pathogenesis and progression. Cancers (Basel). 2010;2(2):483–97.
122. Takeishi S, et al. Ablation of Fbxw7 eliminates leukemia-initiating cells by preventing quiescence. Cancer Cell. 2013;23(3):347–61.
123. Reavie L, et al. Regulation of c-Myc ubiquitination controls chronic myelogenous leukemia initiation and progression. Cancer Cell. 2013;23(3):362–75.
124. Okabe H, et al. A critical role for FBXW8 and MAPK in cyclin D1 degradation and cancer cell proliferation. PLoS One. 2006;1:e128.
125. Kim SJ, et al. mTOR complex 2 regulates proper turnover of insulin receptor substrate-1 via the ubiquitin ligase subunit Fbw8. Mol Cell. 2012;48(6):875–87.
126. Tsunematsu R, et al. Fbxw8 is essential for Cul1-Cul7 complex formation and for placental development. Mol Cell Biol. 2006;26(16):6157–69.
127. Tsutsumi T, et al. Disruption of the Fbxw8 gene results in pre- and postnatal growth retardation in mice. Mol Cell Biol. 2008;28(2):743–51.

128. Kanie T, et al. Genetic reevaluation of the role of F-box proteins in cyclin D1 degradation. Mol Cell Biol. 2012;32(3):590–605.

129. Litterman N, et al. An OBSL1-Cul7Fbxw8 ubiquitin ligase signaling mechanism regulates Golgi morphology and dendrite patterning. PLoS Biol. 2011;9(5):e1001060.

130. Guardavaccaro D, et al. Control of meiotic and mitotic progression by the F box protein beta-Trcp1 in vivo. Dev Cell. 2003;4(6):799–812.

131. Kanarek N, et al. Spermatogenesis rescue in a mouse deficient for the ubiquitin ligase SCF{beta}-TrCP by single substrate depletion. Genes Dev. 2010;24(5):470–7.

132. Chen S, et al. An insertion/deletion polymorphism in the 3′ untranslated region of beta-transducin repeat-containing protein (betaTrCP) is associated with susceptibility for hepato-cellular carcinoma in Chinese. Biochem Biophys Res Commun. 2010;391(1):552–6.

133. Scott DK, et al. Identification and analysis of tumor suppressor loci at chromosome 10q23.3-10q25.3 in medulloblastoma. Cell Cycle. 2006;5(20):2381–9.

134. Matsuoka S, et al. Fbxw7 acts as a critical fail-safe against premature loss of hematopoietic stem cells and development of T-ALL. Genes Dev. 2008;22(8):986–91.

135. Maser RS, et al. Chromosomally unstable mouse tumours have genomic alterations similar to diverse human cancers. Nature. 2007;447(7147):966–71.

136. Agrawal N, et al. Exome sequencing of head and neck squamous cell carcinoma reveals inactivating mutations in NOTCH1. Science. 2011;333(6046):1154–7.

137. Le Gallo M, et al. Exome sequencing of serous endometrial tumors identifies recurrent somatic mutations in chromatin-remodeling and ubiquitin ligase complex genes. Nat Genet. 2012;44(12):1310–5.

138. Huber C, et al. Identification of mutations in CUL7 in 3-M syndrome. Nat Genet. 2005;37(10):1119–24.

139. Baguma-Nibasheka M, Kablar B. Abnormal retinal development in the Btrc null mouse. Dev Dyn. 2009;238(10):2680–7.

140. Matsumoto A, et al. Fbxw7beta resides in the endoplasmic reticulum membrane and protects cells from oxidative stress. Cancer Sci. 2011;102(4):749–55.

141. Jandke A, et al. The F-box protein Fbw7 is required for cerebellar development. Dev Biol. 2011;358(1):201–12.

142. Hoeck JD, et al. Fbw7 controls neural stem cell differentiation and progenitor apoptosis via Notch and c-Jun. Nat Neurosci. 2010;13(11):1365–72.

143. Sancho R, et al. Fbw7 repression by hes5 creates a feedback loop that modulates notch-mediated intestinal and neural stem cell fate decisions. PLoS Biol. 2013;11(6):e1001586.

144. Kwon YW, et al. Pten regulates Aurora-A and cooperates with Fbxw7 in modulating radiation-induced tumor development. Mol Cancer Res. 2012;10(6):834–44.

145. Bhatia N, et al. Role of beta-TrCP ubiquitin ligase receptor in UVB mediated responses in skin. Arch Biochem Biophys. 2011;508(2):178–84.

146. Davis H, et al. FBXW7 mutations typically found in human cancers are distinct from null alleles and disrupt lung development. J Pathol. 2011;224(2):180–9.

147. Kitagawa M, et al. An F-box protein, FWD1, mediates ubiquitin-dependent proteolysis of beta-catenin. EMBO J. 1999;18(9):2401–10.

148. Pan Y, et al. Sonic hedgehog signaling regulates Gli2 transcriptional activity by suppressing its processing and degradation. Mol Cell Biol. 2006;26(9):3365–77.

149. Jia J, et al. Phosphorylation by double-time/CKIepsilon and CKIalpha targets cubitus inter-ruptus for Slimb/beta-TRCP-mediated proteolytic processing. Dev Cell. 2005;9(6):819–30.

150. Wang B, Li Y. Evidence for the direct involvement of {beta}TrCP in Gli3 protein processing. Proc Natl Acad Sci U S A. 2006;103(1):33–8.

151. Zhao B, et al. A coordinated phosphorylation by Lats and CK1 regulates YAP stability through SCF(beta-TRCP). Genes Dev. 2010;24(1):72–85.

152. Liu CY, et al. The hippo tumor pathway promotes TAZ degradation by phosphorylating a phosphodegron and recruiting the SCF{beta}-TrCP E3 ligase. J Biol Chem. 2010;285(48):37159–69.

153. Lee YJ, et al. HSF1 as a mitotic regulator: phosphorylation of HSF1 by Plk1 is essential for mitotic progression. Cancer Res. 2008;68(18):7550–60.

154. Tsuchiya Y, et al. Dual regulation of the transcriptional activity of Nrf1 by beta-TrCP- and Hrd1-dependent degradation mechanisms. Mol Cell Biol. 2011;31(22):4500–12.

155. Rada P, et al. SCF/{beta}-TrCP promotes glycogen synthase kinase 3-dependent degradation of the Nrf2 transcription factor in a Keap1-independent manner. Mol Cell Biol. 2011;31(6): 1121–33.

156. Lassot I, et al. ATF4 degradation relies on a phosphorylation-dependent interaction with the SCF(betaTrCP) ubiquitin ligase. Mol Cell Biol. 2001;21(6):2192–202.

157. Manavathi B, Rayala SK, Kumar R. Phosphorylation-dependent regulation of stability and transforming potential of ETS transcriptional factor ESE-1 by p21-activated kinase 1. J Biol Chem. 2007;282(27):19820–30.

158. Popov N, et al. Ubiquitylation of the amino terminus of Myc by SCF(beta-TrCP) antagonizes SCF(Fbw7)-mediated turnover. Nat Cell Biol. 2010;12(10):973–81.

159. Gallegos JR, et al. SCF TrCP1 activates and ubiquitylates TAp63gamma. J Biol Chem. 2008;283(1):66–75.

160. Guardavaccaro D, et al. Control of chromosome stability by the beta-TrCP-REST-Mad2 axis. Nature. 2008;452(7185):365–9.

161. Westbrook TF, et al. SCFbeta-TRCP controls oncogenic transformation and neural differentiation through REST degradation. Nature. 2008;452(7185):370–4.

162. Yang L, et al. Acute myelogenous leukemia-derived SMAD4 mutations target the protein to ubiquitin-proteasome degradation. Hum Mutat. 2006;27(9):897–905.

163. Soond SM, et al. ERK and the F-box protein betaTRCP target STAT1 for degradation. J Biol Chem. 2008;283(23):16077–83.

164. Spengler ML, Guo LW, Brattain MG. Phosphorylation mediates Sp1 coupled activities of proteolytic processing, desumoylation and degradation. Cell Cycle. 2008;7(5):623–30.

165. Wei S, et al. Thiazolidinediones mimic glucose starvation in facilitating Sp1 degradation through the up-regulation of beta-transducin repeat-containing protein. Mol Pharmacol. 2009;76(1):47–57.

166. Zhong J, et al. Degradation of the transcription factor Twist, an oncoprotein that promotes cancer metastasis. Discov Med. 2013;15(80):7–15.

167. Wu C, Ghosh S. beta-TrCP mediates the signal-induced ubiquitination of IkappaBbeta. J Biol Chem. 1999;274(42):29591–4.

168. Orian A, et al. SCF(beta)(-TrCP) ubiquitin ligase-mediated processing of NF-kappaB p105 requires phosphorylation of its C-terminus by IkappaB kinase. EMBO J. 2000;19(11): 2580–91.

169. Amir RE, Iwai K, Ciechanover A. The NEDD8 pathway is essential for SCF(beta -TrCP)-mediated ubiquitination and processing of the NF-kappa B precursor p105. J Biol Chem. 2002;277(26):23253–9.

170. Lang V, et al. betaTrCP-mediated proteolysis of NF-kappaB1 p105 requires phosphorylation of p105 serines 927 and 932. Mol Cell Biol. 2003;23(1):402–13.

171. Fong A, Sun SC. Genetic evidence for the essential role of beta-transducin repeat-containing protein in the inducible processing of NF-kappa B2/p100. J Biol Chem. 2002;277(25): 22111–4.

172. Xiao G, Fong A, Sun SC. Induction of p100 processing by NF-kappaB-inducing kinase involves docking IkappaB kinase alpha (IKKalpha) to p100 and IKKalpha-mediated phosphorylation. J Biol Chem. 2004;279(29):30099–105.

173. Amir RE, et al. Mechanism of processing of the NF-kappa B2 p100 precursor: identification of the specific polyubiquitin chain-anchoring lysine residue and analysis of the role of NEDD8-modification on the SCF(beta-TrCP) ubiquitin ligase. Oncogene. 2004;23(14):2540–7.

174. Lobry C, et al. Negative feedback loop in T cell activation through IkappaB kinase-induced phosphorylation and degradation of Bcl10. Proc Natl Acad Sci U S A. 2007;104(3):908–13.

175. Shi P, et al. Persistent stimulation with interleukin-17 desensitizes cells through SCFbeta-TrCP-mediated degradation of Act1. Sci Signal. 2011;4(197):ra73.

176. Cui W, et al. beta-TrCP-mediated IRAK1 degradation releases TAK1-TRAF6 from the membrane to the cytosol for TAK1-dependent NF-kappaB activation. Mol Cell Biol. 2012;32(19):3990–4000.
177. Kanemori Y, Uto K, Sagata N. Beta-TrCP recognizes a previously undescribed nonphosphorylated destruction motif in Cdc25A and Cdc25B phosphatases. Proc Natl Acad Sci U S A. 2005;102(18):6279–84.
178. Uchida S, et al. SCFbeta(TrCP) mediates stress-activated MAPK-induced Cdc25B degradation. J Cell Sci. 2011;124(Pt 16):2816–25.
179. Cunha-Ferreira I, et al. The SCF/Slimb ubiquitin ligase limits centrosome amplification through degradation of SAK/PLK4. Curr Biol. 2009;19(1):43–9.
180. Guderian G, et al. Plk4 trans-autophosphorylation regulates centriole number by controlling betaTrCP-mediated degradation. J Cell Sci. 2010;123(Pt 13):2163–9.
181. Seki A, et al. Plk1- and beta-TrCP-dependent degradation of Bora controls mitotic progression. J Cell Biol. 2008;181(1):65–78.
182. Limon-Mortes MC, et al. UV-induced degradation of securin is mediated by SKP1-CUL1-beta TrCP E3 ubiquitin ligase. J Cell Sci. 2008;121(Pt 11):1825–31.
183. Shirogane T, et al. SCFbeta-TRCP controls clock-dependent transcription via casein kinase 1-dependent degradation of the mammalian period-1 (Per1) protein. J Biol Chem. 2005; 280(29):26863–72.
184. Eide EJ, et al. Control of mammalian circadian rhythm by CKIepsilon-regulated proteasome-mediated PER2 degradation. Mol Cell Biol. 2005;25(7):2795–807.
185. Mamely I, et al. Polo-like kinase-1 controls proteasome-dependent degradation of Claspin during checkpoint recovery. Curr Biol. 2006;16(19):1950–5.
186. Peschiaroli A, et al. SCFbetaTrCP-mediated degradation of Claspin regulates recovery from the DNA replication checkpoint response. Mol Cell. 2006;23(3):319–29.
187. Mailand N, et al. Destruction of Claspin by SCFbetaTrCP restrains Chk1 activation and facilitates recovery from genotoxic stress. Mol Cell. 2006;23(3):307–18.
188. Kee Y, Kim JM, D'Andrea AD. Regulated degradation of FANCM in the Fanconi anemia pathway during mitosis. Genes Dev. 2009;23(5):555–60.
189. Zhou X, et al. Targeted polyubiquitylation of RASSF1C by the Mule and SCFbeta-TrCP ligases in response to DNA damage. Biochem J. 2012;441(1):227–36.
190. Li ML, Defren J, Brewer G. Hsp27 and F-box protein beta-TrCP promote degradation of mRNA decay factor AUF1. Mol Cell Biol. 2013;33(11):2315–26.
191. Setoyama D, Yamashita M, Sagata N. Mechanism of degradation of CPEB during Xenopus oocyte maturation. Proc Natl Acad Sci U S A. 2007;104(46):18001–6.
192. Chu PC, et al. The mRNA-stabilizing factor HuR protein is targeted by beta-TrCP protein for degradation in response to glycolysis inhibition. J Biol Chem. 2012;287(52):43639–50.
193. Wang C, et al. The F-box protein beta-TrCP promotes ubiquitination of TRF1 and regulates the ALT-associated PML bodies formation in U2OS cells. Biochem Biophys Res Commun. 2013;434(4):728–34.
194. Sahasrabuddhe AA, et al. betaTrCP regulates BMI1 protein turnover via ubiquitination and degradation. Cell Cycle. 2011;10(8):1322–30.
195. Zou C, et al. LPS impairs phospholipid synthesis by triggering beta-transducin repeat-containing protein (beta-TrCP)-mediated polyubiquitination and degradation of the surfactant enzyme acyl-CoA:lysophosphatidylcholine acyltransferase I (LPCAT1). J Biol Chem. 2011;286(4):2719–27.
196. Tudzarova S, et al. Two ubiquitin ligases, APC/C-Cdh1 and SKP1-CUL1-F (SCF)-beta-TrCP, sequentially regulate glycolysis during the cell cycle. Proc Natl Acad Sci U S A. 2011;108(13):5278–83.
197. Meloche S, Roux PP. F-box proteins elongate translation during stress recovery. Sci Signal. 2012;5(227):e25.
198. Kruiswijk F, et al. Coupled activation and degradation of eEF2K regulates protein synthesis in response to genotoxic stress. Sci Signal. 2012;5(227):ra40.
199. Meyer L, et al. beta-Trcp mediates ubiquitination and degradation of the erythropoietin receptor and controls cell proliferation. Blood. 2007;109(12):5215–22.

200. da Silva Almeida AC, Strous GJ, van Rossum AG. betaTrCP controls GH receptor degradation via two different motifs. Mol Endocrinol. 2012;26(1):165–77.
201. Kumar KG, et al. SCF(HOS) ubiquitin ligase mediates the ligand-induced down-regulation of the interferon-alpha receptor. EMBO J. 2003;22(20):5480–90.
202. Kumar KG, Krolewski JJ, Fuchs SY. Phosphorylation and specific ubiquitin acceptor sites are required for ubiquitination and degradation of the IFNAR1 subunit of type I interferon receptor. J Biol Chem. 2004;279(45):46614–20.
203. Jiang H, et al. Regulation of interleukin-10 receptor ubiquitination and stability by beta-TrCP-containing ubiquitin E3 ligase. PLoS One. 2011;6(11):e27464.
204. Li Y, et al. Negative regulation of prolactin receptor stability and signaling mediated by SCF(beta-TrCP) E3 ubiquitin ligase. Mol Cell Biol. 2004;24(9):4038–48.
205. Shaik S, et al. SCF(beta-TRCP) suppresses angiogenesis and thyroid cancer cell migration by promoting ubiquitination and destruction of VEGF receptor 2. J Exp Med. 2012;209(7):1289–307.
206. Meyer RD, et al. PEST motif serine and tyrosine phosphorylation controls vascular endothelial growth factor receptor 2 stability and downregulation. Mol Cell Biol. 2011;31(10):2010–25.
207. Gao D, et al. mTOR drives its own activation via SCF(betaTrCP)-dependent degradation of the mTOR inhibitor DEPTOR. Mol Cell. 2011;44(2):290–303.
208. Zhao Y, Xiong X, Sun Y. DEPTOR, an mTOR inhibitor, is a physiological substrate of SCF(betaTrCP) E3 ubiquitin ligase and regulates survival and autophagy. Mol Cell. 2011;44(2):304–16.
209. Duan S, et al. mTOR generates an auto-amplification loop by triggering the betaTrCP- and CK1alpha-dependent degradation of DEPTOR. Mol Cell. 2011;44(2):317–24.
210. Li X, Liu J, Gao T. beta-TrCP-mediated ubiquitination and degradation of PHLPP1 are negatively regulated by Akt. Mol Cell Biol. 2009;29(23):6192–205.
211. Asada S, et al. Oxidative stress-induced ubiquitination of RCAN1 mediated by SCFbeta-TrCP ubiquitin ligase. Int J Mol Med. 2008;22(1):95–104.
212. Ang XL, et al. Regulation of postsynaptic RapGAP SPAR by Polo-like kinase 2 and the SCFbeta-TRCP ubiquitin ligase in hippocampal neurons. J Biol Chem. 2008;283(43):29424–32.
213. Zhao J, et al. Extracellular signal-regulated kinase (ERK) regulates cortactin ubiquitination and degradation in lung epithelial cells. J Biol Chem. 2012;287(23):19105–14.
214. Mantovani F, Banks L. Regulation of the discs large tumor suppressor by a phosphorylation-dependent interaction with the beta-TrCP ubiquitin ligase receptor. J Biol Chem. 2003;278(43):42477–86.
215. Hayakawa M, et al. The FWD1/beta-TrCP-mediated degradation pathway establishes a 'turning off switch' of a Cdc42 guanine nucleotide exchange factor, FGD1. Genes Cells. 2005;10(3):241–51.
216. Hayakawa M, et al. Novel insights into FGD3, a putative GEF for Cdc42, that undergoes SCF(FWD1/beta-TrCP)-mediated proteasomal degradation analogous to that of its homologue FGD1 but regulates cell morphology and motility differently from FGD1. Genes Cells. 2008;13(4):329–42.
217. Inuzuka H, et al. Phosphorylation by casein kinase I promotes the turnover of the Mdm2 oncoprotein via the SCF(beta-TRCP) ubiquitin ligase. Cancer Cell. 2010;18(2):147–59.
218. Fukushima H, et al. SCF-mediated Cdh1 degradation defines a negative feedback system that coordinates cell-cycle progression. Cell Rep. 2013;4(4):803–16.
219. Chen H, et al. DNA damage regulates UHRF1 stability via the SCF(beta-TrCP) E3 ligase. Mol Cell Biol. 2013;33(6):1139–48.
220. Peschiaroli A, et al. The ubiquitin-specific protease USP47 is a novel beta-TRCP interactor regulating cell survival. Oncogene. 2010;29(9):1384–93.
221. Douglas JL, et al. Vpu directs the degradation of the human immunodeficiency virus restriction factor BST-2/Tetherin via a {beta}TrCP-dependent mechanism. J Virol. 2009;83(16):7931–47.

222. Mitchell RS, et al. Vpu antagonizes BST-2-mediated restriction of HIV-1 release via beta-TrCP and endo-lysosomal trafficking. PLoS Pathog. 2009;5(5):e1000450.
223. Tian Y, et al. TAZ promotes PC2 degradation through a SCFbeta-Trcp E3 ligase complex. Mol Cell Biol. 2007;27(18):6383–95.
224. Moberg KH, et al. Archipelago regulates Cyclin E levels in Drosophila and is mutated in human cancer cell lines. Nature. 2001;413(6853):311–6.
225. Sundqvist A, et al. Control of lipid metabolism by phosphorylation-dependent degradation of the SREBP family of transcription factors by SCF(Fbw7). Cell Metab. 2005;1(6):379–91.
226. Kanei-Ishii C, et al. Fbxw7 acts as an E3 ubiquitin ligase that targets c-Myb for nemo-like kinase (NLK)-induced degradation. J Biol Chem. 2008;283(45):30540–8.
227. Kitagawa K, et al. Fbw7 promotes ubiquitin-dependent degradation of c-Myb: involvement of GSK3-mediated phosphorylation of Thr-572 in mouse c-Myb. Oncogene. 2009;28(25): 2393–405.
228. Perez-Benavente B, et al. GSK3-SCF(FBXW7) targets JunB for degradation in G2 to preserve chromatid cohesion before anaphase. Oncogene. 2013;32(17):2189–99.
229. Wang R, et al. FBW7 regulates endothelial functions by targeting KLF2 for ubiquitination and degradation. Cell Res. 2013;23(6):803–19.
230. Liu N, et al. The Fbw7/human CDC4 tumor suppressor targets proproliferative factor KLF5 for ubiquitination and degradation through multiple phosphodegron motifs. J Biol Chem. 2010;285(24):18858–67.
231. Bengoechea-Alonso MT, Ericsson J. Tumor suppressor Fbxw7 regulates TGFbeta signaling by targeting TGIF1 for degradation. Oncogene. 2010;29(38):5322–8.
232. Chen MC, et al. Novel mechanism by which histone deacetylase inhibitors facilitate topoisomerase IIalpha degradation in hepatocellular carcinoma cells. Hepatology. 2011;53(1):148–59.
233. Tan M, et al. SAG/RBX2/ROC2 E3 ubiquitin ligase is essential for vascular and neural development by targeting NF1 for degradation. Dev Cell. 2011;21(6):1062–76.
234. Lee JW, et al. The transcription factor STAT2 enhances proteasomal degradation of RCAN1 through the ubiquitin E3 ligase FBW7. Biochem Biophys Res Commun. 2012;420(2): 404–10.
235. Zhao J, et al. FBXW7-mediated degradation of CCDC6 is impaired by ATM during DNA damage response in lung cancer cells. FEBS Lett. 2012;586(24):4257–63.
236. Lochab S, et al. E3 ubiquitin ligase Fbw7 negatively regulates granulocytic differentiation by targeting G-CSFR for degradation. Biochim Biophys Acta. 2013;1833(12):2639–52.
237. Yang CS, et al. FBW2 targets GCMa to the ubiquitin-proteasome degradation system. J Biol Chem. 2005;280(11):10083–90.
238. Werner A, et al. SCFFbxw5 mediates transient degradation of actin remodeller Eps8 to allow proper mitotic progression. Nat Cell Biol. 2013;15(2):179–88.
239. Puklowski A, et al. The SCF-FBXW5 E3-ubiquitin ligase is regulated by PLK4 and targets HsSAS-6 to control centrosome duplication. Nat Cell Biol. 2011;13(8):1004–9.
240. Pagan J, Pagano M. FBXW5 controls centrosome number. Nat Cell Biol. 2011;13(8): 888–90.
241. Kong C, et al. Ubiquitination and degradation of the hominoid-specific oncoprotein TBC1D3 is mediated by CUL7 E3 ligase. PLoS One. 2012;7(9):e46485.
242. Chaturvedi P, Parnaik VK. Lamin A rod domain mutants target heterochromatin protein 1alpha and beta for proteasomal degradation by activation of F-box protein, FBXW10. PLoS One. 2010;5(5):e10620.
243. Zou C, et al. SCF(Fbxw15) mediates histone acetyltransferase binding to origin recognition complex (HBO1) ubiquitin-proteasomal degradation to regulate cell proliferation. J Biol Chem. 2013;288(9):6306–16.

Chapter 3
The Role of FBXL Subfamily
of F-box Proteins in Tumorigenesis

Brian J. North, Yueyong Liu, Hiroyuki Inuzuka, and Wenyi Wei

Abstract The FBXL subfamily is composed of 22 members including the well-characterized FBXL1 (also known as Skp2) and FBXL2 to FBXL21, each containing an F-box motif and a C-terminal Leu-rich repeat (LRR) domain (Fig. 3.1). Intensive studies have revealed an oncogenic role for Skp2, but the potential roles of other FBXL subfamily members in tumorigenesis have just begun to be appreciated. In this chapter, we primarily focus on summarizing the recent genetic, pathological as well as the biochemical evidence pinpointing a possible tumor suppressor or oncogenic role for each of the FBXL subfamily member proteins. In the following paragraphs, we discuss current advances in three major categories, including the physiological (mouse modeling), pathological (human clinical cancer relevance), and biochemical evidence (updated ubiquitin substrates). These three experimental evidence categories will provide insights to facilitate our understanding for their roles in tumorigenesis (Table 3.4). As stated in previous chapters, given the fact that physiological evidence (mouse modeling results) is considered as the strongest supportive data to implicate any given F-box protein in tumorigenesis (Tables 3.2 and 3.3), we choose to summarize FBXL members with available mouse genetic models.

Keywords F-box • SCF • FBXL subfamily • Skp2 • FBXL3 • FBXL10 • FBXL20 • FBXL21 • Cullin 1 • Circadian cycle • Histone demethylase • Tumor suppressor • Oncoprotein • Mouse model • Physiological function

Brian J. North and Yueyong Liu have contributed equally to this chapter.

B.J. North • Y. Liu • H. Inuzuka • W. Wei (✉)
Department of Pathology, Beth Israel Deaconess Medical Center,
Harvard Medical School, Boston, MA 02215, USA
e-mail: wwei2@bidmc.harvard.edu

H. Inuzuka and W. Wei, *SCF and APC E3 Ubiquitin Ligases in Tumorigenesis*,
SpringerBriefs in Cancer Research, DOI 10.1007/978-3-319-05026-3_3,
© The Author(s) 2014

47

3.1 FBXL1 (Skp2) Is an Oncoprotein

FBXL1, also known as Skp2 (S-phase kinase-associated protein 2), is one of the most well-characterized mammalian F-box proteins to date. Skp2 functions as the substrate-recruiting component of the SCF type of E3 ubiquitin ligase complex [1]. Skp2 is a member of the FBXL subfamily [2] and composed of four distinct domains, including a destruction domain (D-box, amino acids 3–6 in the N-terminus) critical for Skp2 stability, a putative nuclear localization signal (NLS, amino acids 66–72), an F-box domain (amino acids 100–150) which mediates the interaction with Skp1 [3], and a C-terminal LRR motif responsible for substrates recognition (Fig. 3.1).

Skp2 was originally identified by Beach and colleagues in 1995 due to its ability to bind Cyclin A. Subsequent studies have demonstrated that Skp2 is a key cell cycle regulator and also functions during many cellular processes that are related to tumorigenesis by targeted degradation of critically important tumor suppressor substrates (Table 3.1) [1]. In this chapter, we highlight its representative substrates, roles of these molecules in tumorigenesis and cancer progression, regulation of Skp2 activity and its overall implications in cancer.

3.1.1 Skp2 Functions as an Oncoprotein that Targets Tumor Suppressors for Degradation

Initially, the tumorigenic role of Skp2 was observed in cell culture systems, in which elevated expression of Skp2 was found to induce degradation of p27, promote entry into S phase, and induce proliferation in the absence of adhesion to the extracellular matrix in immortalized cells [4]. Ectopic expression of a dominant-negative version of Skp2 in breast cancer cells led to reduced colony formation in soft agar [5]. In addition, androgen signaling was reported to promote the expression of Skp2 in human prostate cells, leading to an increase in p27 degradation and proliferation [6, 7]. Subsequently, relatively higher Skp2 expression has been frequently detected in various types of cancers, including lymphomas [8, 9], pancreatic cancer [10], breast carcinomas [3, 11–14], prostate cancer [15, 16], melanoma [17–19], and nasopharyngeal carcinoma [20, 21]. Moreover, gene amplification of *Skp2* correlates with poor prognosis in human gastric cancer [22]. Recently, it has been reported that Skp2 expression correlated significantly with histological grade and tumor size in human hepatocarcinoma [23]. In line with these findings in human cancers, *Skp2* knockout mice are resistant to tumor development induced by loss of either *p19^Arf* or the *Pten* tumor suppressor protein [24] (Table 3.2). Furthermore, overexpression of Skp2B, an isoform of Skp2 with a varied carboxyl terminal domain, induced tumor proliferation and growth in a xenograft mouse model [25]. Taken together, these results indicate that Skp2 might function as a bona fide oncoprotein.

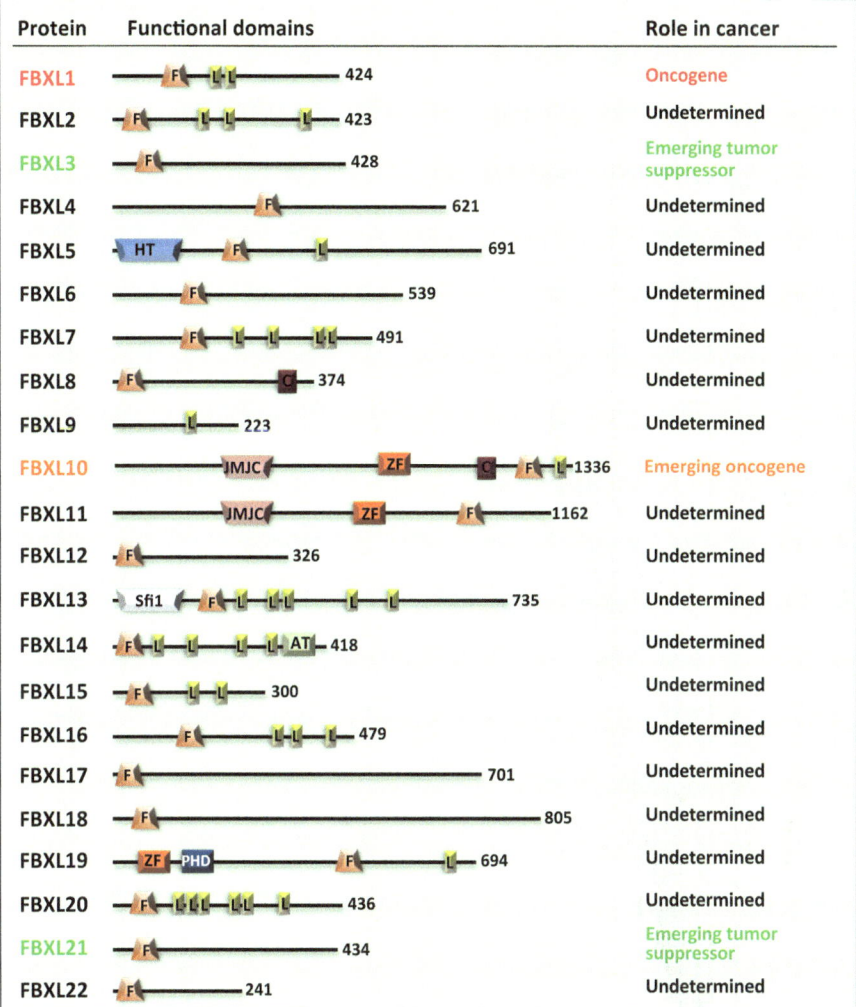

Fig. 3.1 A schematic illustration of functional domains of all known FBXL proteins and their possible roles in cancer. *F* F-box motif, *L* leucine-rich repeat, *JMJC* domain found in cupin metalloenzyme family, *ZF* CXXC zinc finger domain, *Sfi1* Sfi1 spindle body protein, *AT* antitermination protein domain, *PHD* plant homeo domain

Notably, *Skp2*$^{-/-}$ mice show elevated levels of p27 protein suggesting that p27 is a primary target of Skp2 [26, 27] (Table 3.2). In further support of this notion, expression of Skp2 is inversely related to p27 expression during the differentiation of human embryonic stem cells and in many human tumors [28–30]. *Skp2*$^{-/-}$ MEFs displayed reduced cell proliferation, accompanied by increased p27 protein

Table 3.1 Summary of the identified ubiquitin substrates for Skp2

Substrates	Functions/signaling pathways	Phospho-degron/ phosphorylation sites	References
p27	Cdk inhibitor, Cell cycle	Thr187	[24, 127, 133]
p21	Cdk inhibitor, Cell cycle	Ser130	[71, 128]
p57	Cdk inhibitor, Cell cycle	Thr310	[129]
p130	Rb protein family, Cell cycle	Ser672	[126, 130]
Cyclin A	Cyclin, Cell cycle	–	[120]
Cyclin E	Cyclin, Cell cycle	–	[120]
Cyclin D1	Cyclin, Cell cycle	–	[32]
Cyclin G2	Cyclin, Cell cycle	–	[68]
Cdt1	DNA replication factor, Cell cycle	–	[31]
Orc1	Component of origin recognition complex, Cell cycle		[135]
E2A	Transcription factor, Muscle or early B-cell differentiation	–	[136]
c-Myc	Transcription factor, Cell proliferation		[117, 137]
Smad4	Transcription factor, TGFβ signaling pathway	–	[138]
FOXO1	Transcription factor, Apoptosis and cell metabolism	Ser256	[139]
FOXO3	Transcription factor, Apoptosis and cell metabolism	–	[111]
MEF	Transcription factor, Natural killer T cell proliferation and innate immunity	Ser641, Thr643, Thr648	[140]
Tob1	ErbB2 interacting protein, Inhibition of cell proliferation	–	[141]
MLL	Histone methyltransferase, Chromatin regulation	–	[117]
ERα	Transcription factor, Nuclear receptor signaling	Ser294	[142]
HPV-E7	HPV oncoprotein, Cellular transformation	–	[143]
HPV18-E2	HPV oncoprotein, Cellular transformation	–	[144]
USP18	Deubiquitinating protease, Ubiquitin conjugation pathway	–	[145]
BRCA2	DNA repairing protein, Cell cycle and DNA damage	–	[146]
β-TRCP	F-box protein, Ubiquitin conjugation pathway	Ser412	[147]
RAG2	RAG component, V(D)J recombination	Thr490	[148]
DUSP1	Protein phosphatase, MAPK signaling pathway	Ser296, Ser323	[142]
BTG2	Transcription co-regulator, Cell cycle	–	[149]
RASSF1A	Ras interacting protein, Cell cycle	Ser203	[150]
ING3	Transcription co-regulator, Inhibition of cell proliferation	–	[151]
Prohibitin	Transcription co-regulator, Cell cycle	–	[152]
E-cadherin	Cell adhesion protein, Cell adhesion	Ser840, Ser842	[152]

(continued)

Table 3.1 (continued)

Substrates	Functions/signaling pathways	Phospho-degron/ phosphorylation sites	References
SETD8	Histone Methyltransferase, Chromatin regulation	–	[153]
JAK3	Protein kinase, Cytokine signaling pathway	–	[140]
SIRT3	NAD-dependent protein deacetylase, Metabolic process	–	[154, 155]
Akt1	Protein kinase, PI3K signaling	–	[156]
NBS1	Component of MRE11/RAD50/NBS1 complex, DNA damage response	–	[157]
B-Myb	Transcription factor, Cell cycle	–	[158]
E2F1	Transcription factor, Cell cycle	–	[159]
Tal-1	Transcription factor, Erythrocyte maturation	Ser300	[160]

Table 3.2 Summary of the knockout mouse models for the FBXL class of F-Box proteins

F-box protein	Knockout/phenotype
FBXL1 (SKP2) Oncogene	*Skp2−/−*: Smaller littermates [31] Inhibited skin tumorigenesis [69] Renal injury [126] Decreased functional gamete reserve [127] Smaller neointimal areas [128] Diminished beta cell mass, hypoinsulinemia, and glucose intolerance [129] Protected mice from the development of obesity [130] Massive apoptosis [31] in spermatogenic cells [131] Hepatocytes entered the endoduplication cycle after mitogenic stimulation [32] Hepatocytes increased in size [66] *Compound mice models* *Pten+/−/Skp2−/− and Arf−/−/Skp2−/−*: senescence [24] *Rb+/−/Skp2−/−*: no tumor [71] *Skp2−/−/p27−/−*: skin tumor [69], progression of nephropathy [27] *Pten+/−/Skp2−/−*: normal HSC quiescence [132]
FBXL3 Emerging oncogene	*Fbxl3−/−*: long circadian period [117]
FBXL10 Emerging oncogene	*Fbxl10−/−*: half pups died after birth [125]
FBXL20 Undetermined	*Fbxl20−/−*: long-term potentiation/depression [112] *Fbxl20+/−*: defect fear memory [111]
FBXL21 Emerging tumor suppressor	*Fbxl21−/−*: compromised organization of daily activities [117]

Table 3.3 Summary of the transgenic mouse models for the FBXL class of F-Box proteins

| | Transgenic mouse model/phenotype | | | | |
| | Whole-body expression/phenotype | | Tissue-specific expression | | |
F-box protein	Transgene	Phenotype	Tissue	Transgene	Phenotype
FBXL1 (SKP2) Oncogene	BCR-ABL-infected *Skp2^-/-* marrow	Myeloproliferative syndrome [65]	B cells	SKP2	No tumor [64]
	p27T187A knock-in in *Skp2^-/-*	Short period in G1 [120]	Mammary glands	MMTV-Skp2B	Breast tumor [133]
	K5-Myc-*Skp2^-/-*	Oral epithelium tumorigenesis [134]	Prostate	Skp2	Prostate cancer [68]
			T-lymphoid	Skp2 and/or NRas	T cell lympho-mas [121]
FBXL20 Undetermined			Hippocampus	Scrapper	No significant difference [111]

expression [31]. Conversely, double deficiency of *p27* and *Skp2* rescued the cell proliferation defects in *Skp2^-/-* MEFs [32], further suggesting that p27 is a critical and physiological Skp2 substrate. Moreover, in vivo evidence showing that *p27* and *Skp2* double deficiency in mice rescued the reduced organ size and body weight observed in *Skp2^-/-* mice [32]. Direct evidence supporting a role for Skp2 in regulating p27 was demonstrated through biochemical analysis where Skp2 ubiquitylated the tumor suppressor p27 in vitro and in vivo [33, 34]. In particular, Skp2 recognized specifically p27 phosphorylated on Thr187, which is catalyzed by various Cdk complexes. Moreover, ubiquitination of p27 required CKS1 (Cdk subunit 1), which binds to Skp2 and increases the affinity of phosphorylated p27 for Skp2. More importantly, *CKS1* deficiency prevents Skp2 binding to p27, in turn leading to increased p27 protein stability [35, 36]. Noteworthy, this result was the first demonstration of a requirement of an accessory protein for SCF function.

In addition to p27, Skp2 also regulates the ubiquitination and subsequent degradation of many other tumor suppressor proteins including p21 [37], p57 [38], TOB1 [39], RASSF1 (Ras association domain family 1) [40], and RBL2 (retinoblastoma-like 2; also known as p130) [41] (Table 3.3). Additionally, FOXO1, a member of the forkhead box-containing transcription factors which is involved in various cellular processes including cell cycle regulation, differentiation, stress responses, and apoptosis is also targeted for degradation by Skp2 in a phosphorylation-specific manner where the degron is targeted by the oncogenic kinase Akt [42, 43].

3.1.2   Regulation of Skp2 in Tumor Progression

3.1.2.1   Transcriptional Regulation of Skp2

Recent genetic and bioinformatic analyses have revealed that Skp2 was positively regulated at the transcriptional level by several transcription factors, including HBXIP [44], E2F1 [45], NF-κB [46], SP1 [47], CBF1 [48], GABP (GA-binding protein) [49], STAT3 [50, 51], and FoxM1 [52]. Studies using overexpression or depletion of these transcription factors often led to the corresponding changes in Skp2 protein levels and downstream effects in regulation of target substrate degradation such as p27. Many of these transcription factors show enhanced activities in cancer, leading to elevated Skp2 abundance, which may facilitate tumorigenesis in the absence of Skp2 mutations directly affecting its abundance or activity.

On the other hand, transcriptional suppression of *Skp2* has also been reported, although the molecular mechanisms still remain largely elusive. A recent report identified Foxp3 as a novel transcriptional repressor for *Skp2* in breast cancer [53]. Reduction of Skp2 by Foxp3 resulted in the accumulation of p27 and induced cell cycle arrest. Interestingly, Skp2 overexpression has been observed in human breast cancer samples and is correlated with Foxp3 downregulation, raising the possibility that Skp2 upregulation may contribute to the development of breast cancer in humans, which is also observed in *Foxp3*$^{+/-}$ mice [54].

3.1.2.2   Regulation of Skp2 by Upstream Signaling Pathways

Although in recent years a large body of evidence implicating Skp2 and its oncogenic roles has been presented, the upstream regulators of Skp2 in human cancer progression still remain poorly defined. Several groups have found that multiple genes and signaling pathways can regulate Skp2 expression.

As a vital regulator of cell cycle progression, Skp2 expression is tightly and timely controlled by proteasome-mediated degradation. Specifically, the APC^{Cdh1} E3 ligase complex plays a primary role in controlling Skp2 levels during the G1 phase, in which APC^{Cdh1} triggers Skp2 ubiquitination and subsequent degradation by the 26S proteasome [55, 56]. An N-terminal D-box motif in Skp2 is responsible for Cdh1 binding, thus and deletion or mutation of this motif resulted in resistance of Skp2 to Cdh1-mediated ubiquitination and degradation [56]. Consistent with a role in regulating Skp2 stability, depletion of Cdh1 led to accumulation of Skp2, which in turn promoted S-phase transition. This is further in line with the fact that Cdh1 protein levels are low during the G1 to S transition, where Skp2 expression reaches its peak [57].

Several reports have recently demonstrated that Skp2 is also regulated by microRNAs. The microRNA miR-7 acts as a tumor suppressor in many cancers [58], where upregulation of miR-7 induces transient cell cycle arrest in the G1 phase without promoting apoptosis by increasing levels of p27 through downregulation of Skp2 [59]. In addition, Skp2 is also a target of miR-203, and downregulation of

Skp2 through this mechanism may be involved in the suppression of self-renewal during epidermal differentiation [60]. However, the physiological contribution of microRNA-mediated regulation of Skp2 during tumorigenesis still remains largely unknown. Therefore, assessing the levels of Skp2 and miR-7/miR-203 in human tumors would be an interesting starting point for future clinical cancer studies.

Furthermore, Androgen Receptor (AR), a ligand-activated transcription factor, has been well documented to play a critical role in the pathological progression of human cancer. Emerging evidence has shown that androgen and AR are involved in regulation of Skp2 [61]. A recent study has shown that AR regulates Skp2 through blocking its D-box-dependent degradation [62]. Pernicova and colleagues found that androgen depletion decreased prostate cancer cell proliferation partly through downregulation of Skp2. Specifically, androgen represses Skp2 expression via both p107-dependent and p107-independent pathways in prostate cancer cells [62, 63]. These results suggest that targeting Skp2 could be a potent therapy for androgen receptor-positive castration-resistant prostate cancer. However, additional studies are required to validate this notion.

Consistent with its oncogenic role, Skp2 also plays critical roles in cancer metastasis. A recent report found that cross talk between TGFβ and Skp2 function to induce the epithelial–mesenchymal transition (EMT). Specifically, TGF-β1 treatment increased Skp2 expression accompanied with increased phosphorylation of Akt1, and accumulation of c-Myc protein, during EMT [18]. Moreover, c-Myc and Akt1, both of which are positive regulators of metastasis, are required for TGF-β1-mediated Skp2 upregulation. The c-Myc transcription factor specifically binds to the promoter of *Skp2* and enhances *Skp2* transcription, while phosphorylation of Skp2 at Ser72 by Akt also attenuates its ubiquitination and degradation [18]. Taken together, these reports suggest that Skp2 synergizes with TGF-β, Akt, and c-Myc to regulate metastasis in human cancers.

3.1.3   Skp2 Knockout Mouse Models: Implication for Tumor Development

To better understand the underlying mechanisms of the oncogenic function of Skp2, various transgenic and knockout mouse models have been developed (Tables 3.2 and 3.3). Induction of cellular apoptosis, senescence, and inhibition of tumor development have been observed in knockout mouse studies, whereas induction of tumorigenesis and shorter latency were observed in transgenic mouse models. These observations account, in part, for the tumorigenesis process in T-cell lineage [31], B-cell lineage [64], bone marrow [65], liver [32, 66], breast [67], prostate [68], and skin [69].

The Nakayama group generated the first *Skp2*$^{-/-}$ mouse [31] which were viable but with smaller size and reduced body weight. Isolated cells from these mice exhibited markedly enlarged nuclei with polyploidy and multiple centrosomes, and showed a reduced growth rate and increased apoptosis. Consistently, human cells depleted of Skp2 also exhibit increased accumulation of both Cyclin E and p27. The elimination of Cyclin E during S and G2 phases is impaired in $Skp2^{-/-}$ cells, resulting in loss of Cyclin E periodicity [31]. Biochemical studies found that Skp2 interacts specifically with Cyclin E and thereby promotes its ubiquitination and degradation both in vivo and in vitro [13, 70]. Similarly, loss of $Skp2$ resulted in p27 accumulation as well as enlargement and polyploidy of hepatocytes, thereby impairing hepatocyte proliferation, which is compensated for by cellular enlargement during liver regeneration [66]. Hepatocytes of $Skp2^{-/-}$ mice enter the endoduplication cycle after mitogenic stimulation, whereas this phenotype was not apparent in $Skp2^{-/-}/p27^{-/-}$ mice, suggesting that aberrant increase in p27 stability might be necessary for the activation of Cdk2, and that Skp2 contributes to regulation of G2 to M progression by mediating the degradation of p27 [32]. In line with these findings, inactivation of $Skp2$ in $Rb^{+/-}$ mice completely blocked tumor development caused by Rb depletion. Correspondingly, knock-in mice with the $p27$-$Thr187A$ mutant reproduced the effects of $Skp2$ knockout in $Rb^{+/-}$ mice, confirming that p27 ubiquitination by SCF^{Skp2} ubiquitin ligase might be the primary mechanism underlying the essential role of Skp2 in tumorigenesis [71]. Notably, mice that received transplants of BCR-ABL-infected $Skp2^{-/-}$ bone marrow developed a myeloproliferative syndrome but survival was significantly prolonged compared with recipients of BCR-ABL-expressing $Skp2^{+/+}$ bone marrow [65]. Lin et al. further found that Skp2 inactivation restricted tumorigenesis following loss of Arf by crossing $Skp2^{-/-}$ with $Arf^{-/-}$ mice, where Skp2 deletion markedly prolonged the overall survival of $Arf^{-/-}$ mice. Approximately 33 % of $Arf^{-/-}$ mice developed sarcoma and/or lymphoma within 1 year whereas $Skp2^{-/-}/Arf^{-/-}$ compound mutant mice showed no signs of tumor formation [24]. These results pinpoint Skp2 inhibition as a viable approach to cancer treatment.

On the other hand, overexpression of Skp2 in mice significantly promotes tumor development exhibiting a short latency period and higher penetrance (Table 3.3). Pagano observed that Skp2 expression correlated positively with tumor grade of malignancy, and inversely with p27 levels in human lymphomas. Furthermore, transgenic mice expressing Skp2 targeted to the T-lymphoid lineage as well as double transgenic mice co-expressing Skp2 and activated N-Ras led to the development of lymphoma [72]. A strong cooperative effect between these two transgenes led to a significant decrease in survival when compared with control and single transgenic animals. Furthermore, lymphomas of N-Ras single transgenic animals often expressed higher levels of endogenous Skp2 than tumors of double transgenic mice. This study provides supporting evidence of Skp2 as a proto-oncogene that is causally involved in the pathogenesis of lymphomas. Transgenic mouse lines that specifically expressed Skp2 in the prostate gland have also been established [68]. In this transgenic mouse model, aberrant expression of Skp2 was reported to promote marked overproliferation, resulting in hyperplasia, dysplasia, and low-grade carcinoma in the prostate gland. Consistent with its critical role in p27 proteolysis, expression of Skp2 caused significant downregulation of p27 in prostate tissue [68].

Furthermore, Skp2B overexpressing mice showed reduced p53 activity in the mammary gland, which could potentially drive mammary gland tumor development [67]. Taken together, various mouse model studies of Skp2 loss and overexpression have revealed that Skp2 functions as an oncoprotein, and deregulation of Skp2 can promote tumorigenesis through downregulating critical tumor suppressors including p27 and p53.

3.1.4   Skp2 as a Potential Therapeutic Target for Cancer

Given that upregulation of Skp2 induces cancer development in mouse models and overexpression of Skp2 is observed in a variety of human cancers, and depletion of Skp2 in either mouse models or human cancer cell lines shows a reduced growth rate and increased apoptosis; targeting Skp2 can be a potentially effective strategy for cancer therapy. Indeed, large strides have been made and several small molecule inhibitors [73] that can effectively block Skp2 activity have been developed through high-throughput screening [74, 75]. Using in silico screens targeted to the binding interface for p27, Wu et al. identified small molecule inhibitors that specifically inhibited SCF^{Skp2} activity. The advantage of these inhibitors over others is their unique specificity for Skp2–p27 interaction without off-target binding/inhibition of either Cdc34 or Skp1 [76]. Most recently, a small molecule inhibitor Glucosamine, an amino sugar and a prominent precursor in the biochemical synthesis of glycosylated proteins and lipids, has been reported to decrease the expression level of Skp2 and Cyclin E in lung cancer cells [77]. Moreover, inhibition of cell proliferation by blocking the G1 to S transition was also observed, as was downregulation of the levels of phosphorylated Cyclin E (Thr62) and p27Kip1 (Thr187). Furthermore, another small molecule, Salinomycin, inhibited STAT3 activity and thus decreased expression of Stat3-target genes including Skp2 [78]. Salinomycin was found to induce degradation of Skp2, accompanied by p27 accumulation, cell cycle arrest, and apoptosis. Additional inhibitors were identified targeting Skp2 activity using high-throughput in silico screening of large and diverse chemical libraries [79]. Compound #25 was isolated and completely prevented the Skp2–Skp1 interaction, thereby selectively inhibiting SCF^{Skp2} E3 ubiquitin ligase activity but not affecting the activity of other SCF complexes. Notably, inactivation of Skp2 by these inhibitors restricted cancer stem cell traits and cancer progression. In addition, they also found that Herceptin treatment sensitized the Her2-overexpressing tumors with silenced Skp2, indicating that inhibiting Skp2 could benefit the treatments of Her2-positive cancers [80]. Additionally, several natural compounds have been found to inhibit expression of Skp2 in human cancers [42] including curcumin, quercetin, lycopene, silibinin, epigallocatechin-3-gallate, and vitamin D3. The identification of these small molecule Skp2 regulators provide strong evidence in support of the idea that Skp2 could be a potential and promising target to treat human cancer.

3.2   The Emerging Oncogenic Role of FBXL3

FBXL3, a member of the F-box protein family containing leucine-rich repeats, which together with Skp1 and Cullin 1, combine to form a functional SCF E3 ubiquitin ligase complex. Utilizing a ENU-induced mutagenesis screen in mice, a mutation in *Fbxl3* (termed *Afterhours*) was identified which exhibited a long-circadian period, implicating FBXL3 as a regulator of circadian rhythm [81, 82]. The circadian clock is a molecular delayed feedback mechanism modulating the 24-h rhythm cycle directing daily changes in metabolic, physiologic and behavior functions. The transcription factors CLOCK/BMAL1 and NPAS2/BMAL1 largely control the expression of diurnal genes that in turn regulate a variety of downstream effector pathways to elicit physiological changes on a daily basis. In addition, these transcription complexes control the expression of their own inhibitors, the Period (Per1/Per2) and Cryptochrome (Cry1/Cry2) proteins. The Per and Cry proteins then transit back into the nucleus and block transcription of CLOCK/BMAL1 and NPAS/BMAL1, thereby completing a negative feedback loop. To drive clock oscillations, mammalian Cry proteins associate with the Per proteins and together inhibit the transcription of their own genes by the regulation of *Clock/Bmal1* [83].

Along the circadian rhythm pathway, FBXL3 targets Cry1 and Cry2 for degradation, thus regulating Clock/Bmal1-mediated gene expression [81, 82, 84] (Tables 3.4 and 3.5). FBXL3 binds CRY2 by simultaneously interacting with its FAD-binding pocket with a conserved carboxy-terminal tail and occluding its PER-binding domain [85]. This novel F-box-protein–substrate interaction is susceptible to disruption by both FAD and Per proteins, suggesting a novel mechanism for pharmacological targeting this complex and a unique regulatory mechanism of CRY ubiquitination [85]. Recently, the ubiquitin-specific protease 2 (USP2) has been shown to deubiquitinate Cry1, thus opposing the activity of SCF^FBXL3 [86]. It has further been identified that disruption of the Rev-Erbα gene in *Fbxl3*-deficient mice rescued its long-circadian period phenotype, which is likely due to the role of FBXL3 in regulating Rev-Erb/retinoic acid receptor-related orphan receptor-binding element (RRE)-mediated transcription by inactivating the Rev-Erbα:histone deacetylase 3 corepressor complex [87] (Table 3.2).

Alterations in circadian rhythm, specifically those with extended circadian periods, often lead to altered mood and anxiety, as well as result in increased prevalence of biopolar disorder [88–91]. The FBXL3 *Afterhours* exhibited reduced anxiety- and depression-like behavior and evidence of increased locomotor activity [92]. Furthermore, a genome-wide association between variants in FBXL3 and bipolar disorder in human subjects was identified [92], suggesting that FBXL3 might be a critical regulator of bipolar disorders and indicating that identifying positive regulators for FBXL3 might provide novel therapeutic options.

With respect to cancer, *FBXL3* was found to be mutated in 45 % of cell lines derived from high microsatellite instability colorectal cancers [93]. Recent results suggest that FBXL3 may function as an oncoprotein. Expression of Cry1 and Cry2 in glioma tissue was found to be significantly lower than in the surrounding normal

Table 3.4 Summary of the roles of the FBXL class of F-box proteins in cancer

Potential role in cancer (tumor suppressor or oncogene)	F-box protein	Gene symbol	Physiological evidence (mouse models)		Pathological evidence in cancer		Biochemical evidence (major substrates)
			Knockout	Transgenic	Overexpression	Mutation/deletion	
Oncogene	FBXL1 (SKP2)	*SKP2, FBXL1*	Yes [24, 31, 120]	Yes [68, 121]	Yes [5, 122, 123]		p27, p21, p130, FOXO1
Emerging oncogene	FBXL3	*FBXL3, Fbl3, FBXL3A*	Yes [117]			Yes [124]	CRY
Emerging oncogene	FBXL10	*FBXL10*	Yes [125]		Yes [109]		Ink4a, Ink4b, Ink4c, H2A, c-Fos
Undetermined	FBXL20	*FBXL20, Fbl2*	Yes [112]		Yes [114]		N/A
Emerging tumor suppressor	FBXL21	*FBXL21, FBXL3B, Fbl3B*	Yes [117]				CRY

Table 3.5 Summary of identified ubiquitin substrates for the FBXL class of F-box proteins

F-box protein	Substrate	Signaling pathway/functions	References
FBXL2 Undetermined	p85β	The PIK regulatory subunit	[161, 162]
	Amyloid precursor protein (APP)		[135]
	Cyclin D2	Cell cycle	[163]
	Cyclin D3	Cell cycle	[164, 165]
FBXL3 Emerging oncogene	CRY	Cryptochrome, circadian clock	[81, 82, 84, 85, 116, 117, 166]
FBXL4 Undetermined	KDM4A/JMJD2A	Histone lysine demethylase	[167]
FBXL5 Undetermined	p150(Glued)	Binding to dynein and microtubules	[83]
	IRP2	Maintenance of an appropriate intracellular concentration of iron	[168]
FBXL7 Undetermined	Aurora A	Mitotic spindle formation and chromosome segregation	[169]
FBXL10 Emerging oncogene	H2A	Polycomb repressive complex 1(PRC1)	[104]
FBXL12 Undetermined	Ku80	Nonhomologous end joining (NHEJ) double-strand break (DSB) repair pathway	[170]
	Calmodulin kinase I	Calcium-triggered signaling cascade	[171]
FBXL14 Undetermined	Snail1	Master regulator of epithelial to mesenchymal transition	[172]
	SLUG	Critical regulator of neural crest development and invasive behavior during tumor progression	[173]
	Mkp3	Vertebrate embryonic DV axis formation	[174]
	Twist	Embryo development and tumor metastasis	[175]
FBXL15 Undetermined	Smurf1	Regulation of the bone morphogenetic protein (BMP) signaling pathway	[176]
FBXL19 Undetermined	ST2L	Interleukin 33 (IL-33) mediates pulmonary inflammation and immune system-related disorders	[177]
	Rac1	Cytoskeletal reorganization, membrane trafficking	[178]
	RhoA	The assembly of focal adhesions and actin stress fibers	[179]
FBXL21 Emerging tumor suppressor	CRYs	Circadian cycle	[117]
FBXL22 Undetermined	ACTN FLNC	Inherited cardiac and skeletal muscle diseases	[180]

tissues [94]. These results indicate that disturbances in Cry1 and Cry2 protein stability by FBXL3 may disrupt normal circadian rhythm, thus benefiting the survival of glioma cells [94]. Hence, targeting the expression patterns of circadian clock genes in glioma and non-glioma cells may serve as a therapeutic avenue for brain cancer.

A direct link between genes involved in the core regulation of the circadian clock and tumorigenesis has been demonstrated for a variety of cancer types [95]. The rhythm of the circadian clock and cancer are interlinked as cell cycle genes are affected by the molecular components of the circadian clock; including c-Myc, Wee1, Cyclin D, and p21 [96–98]. For instance, activation of Per2 leads to c-Myc overexpression and an increased tumor incidence. Mice with mutations in Cry1 or Cry2 are arrhythmic (lack a circadian rhythm) and have a faster rate of growth of implanted tumors [99]. Epidemiologically, recent data suggest a correlation between cancer and disruption of circadian rhythm as "The Nurses' Health Study" identified that women working rotational night shifts have an increased incidence of breast cancer [100].

Further evidence for a potential link between FBXL3 and cancer through the regulation of Cry1 and Cry2 protein stability is suggested based on the findings that mutations of cryptochromes in $p53^{-/-}$ mice causes a reduction in age adjusted tumor incidence [101]. Triple knockout mice ($p53^{-/-}Cry1^{-/-}Cry2^{+/-}$ or $p53^{-/-}Cry1^{+/-}Cry2^{-/-}$) had delayed death from lymphoblastic lymphoma that usually occurs in $p53^{-/-}$ mice [101]. However, further studies of Cry1/2 levels in a wild-type genetic background with either altered expression or mutations in *FBXL3* would be required to understand this potential connection further.

In an interesting twist to the regulation of circadian rhythm, Cry1 interacts with Tim (Timeless), which is involved in the DNA damage response, potentially through its association with Chk1 and the ATR-ATRIP [102]. Therefore, FBXL3, by regulating Cry1, may influence a number of facets important for tumorigenesis, which may or may not be influenced by dysfunction in circadian rhythm. Although a correlation between disruption of circadian rhythm and cancer exists, and FBXL3 might protect cells from tumorigenesis through the regulation of circadian rhythm, it cannot be excluded that additional yet to be identified substrates of FBXL3-mediated degradation could influence cancer progression. Therefore, future studies in FBXL3 should focus on the identification of novel substrates for SCF[FBXL3]-mediated degradation.

3.3   The Emerging Oncogenic Role of FBXL10

FBXL10 (also referred to as Ndy1, JHDM1B, and KDM2B) is an evolutionarily conserved and ubiquitously expressed bipartite protein containing both an F-box domain and a JmiC domain that carries out demethylase activity [2, 103]. This imparts FBXL10 with dual activities providing intriguing possibilities for biological function. Along these lines, FBXL10 has been shown to have both histone H3K36 demethylase and H2AK119 ubiquitination activities [104] (Tables 3.4 and 3.5), although a direct biological outcome by coregulation of these modifications by FBXL10 has yet to be described.

FBXL10 has been shown to associate with the BCoR complex [105, 106]. BCoR is a transcriptional corepressor which complexes with histone deacetylases and the sequence-specific transcription factor Bcl-6 [107]. The recruitment of an E3 ubiquitin ligase and a histone demethylase by BCoR suggests that BCoR could use a unique combination of epigenetic modifications (acetylation, methylation, and ubiquitination) to direct gene silencing [105]. FBXL10 was also found to function as a binding partner of Ring1B/Rnf2, a RING finger protein member of the Polycomb group (PcG) of proteins. In this manner FBXL10 is associated with a complex that also includes Bcl6 corepressor (BcoR), CK2α, Skp1, and Nspc1/Pcgf1 [106]. Polycomb repressive complex 1 (PRC1) catalyzes lysine 119 monoubiquitylation on H2A (H2AK119ub1) and regulates pluripotency in embryonic stem cells (ESCs) [104]. Recent results identified an association between FBXL10 and Ring1B/Nspc1, forming a noncanonical PRC1, which is required for H2AK119Ub in mouse ESCs. Genome-wide analyses revealed that FBXL10 preferentially binds to CpG islands and colocalizes with Ring1B on Polycomb target genes [104]. Notably, FBXL10 depletion causes a decrease in Ring1B binding to target genes and a major loss of H2AK119Ub, indicating that FBXL10 is involved in both recruitment of this complex as well as potentially providing ubiquitination activity [104]. Consistent with this possibility, genetic analyses revealed that FBXL10 DNA binding capability and integration into PRC1 are required for H2AK119 ubiquitylation. ESCs lacking FBXL10, similar to previously characterized Polycomb mutants, cannot differentiate properly. These results demonstrate that FBXL10 has a key role in regulating Ring1B recruitment to its target genes and H2AK119 ubiquitylation in ESCs [104]. The putative enzymatic activities and protein interaction and chromatin binding motifs present in this novel Ring1B–Fbxl10 complex suggest additional mechanisms for chromatin recruitment/modification and a possibility for Ring1B/Rnf2 activities beyond those typically associated with PcG function [106], and therefore shedding new light on ubiquitination in regulation of PcG mediated transcriptional regulation.

Our understanding of a direct role for FBXL10 in tumorigenesis has begun to take shape. Using a genome-wide RNAi screen, the JHDM1B worm orthologue (T26A5.5) was identified as a gene that regulates growth in *C. elegans* [108]. In the mouse, four independent screens have identified FBXL10 as a putative tumor suppressor by retroviral insertion analysis [108]. Through its JmiC demethylase domain, human FBXL10 demethylates trimethylated lysine 4 on histone H3 and represses the transcription of ribosomal RNA genes, thereby regulating cell size through translational control mechanisms. Dysregulation of ribosome biogenesis contributes to cellular transformation, and consistent with this, FBXL10 expression is suppressed in aggressive brain tumors, providing compelling evidence of an important role for FBXL10 in tumorigenesis [108].

Direct evidence for dysregulation of FBXL10 in human tumors has recently been identified. FBXL10 is markedly overexpressed in human pancreatic ductal adenocarcinoma (PDAC), one of the most lethal forms of human cancers [109].

Furthermore, levels of FBXL10 correlate with disease grade and stage, with highest expression observed in metastatic PDAC [109]. Depletion of FBXL10 reduced, and overexpression increased, tumorigenesis in PDAC cell lines and in mouse models, respectively [109]. FBXL10 was also shown to be involved in two transcriptional mechanisms to drive tumorigenesis. First, FBXL10 binds to PcG proteins and represses developmental genes. Second, FBXL10 activates a cluster of metabolic genes, including mediators of protein synthesis and mitochondrial function, through association with the MYC oncogene and the histone demethylase KDM5A [109].

In light of these results, and the dual enzymatic nature of the FBXL10 protein, it would be important to further assess the specific role of the putative E3 ubiquitin ligase activity of FBXL10 in tumorigenesis, and how ubiquitination and demethylation, carried about by FBXL10, is coordinately regulated during tumorigenesis.

3.4   The Role of FBXL20 in Tumorigenesis Remains Largely Undetermined

FBXL20, an F-box protein also known as SCRAPPER, is a subunit of SCF type E3 ubiquitin ligase [110]. FBXL20 is found in synapses where it binds to and mediates the ubiquitination of Rab3-interacting molecule 1 (RIM1), an essential factor for synaptic vesicle release, implicating FBXL20 in regulation of neural transmission and synaptic tuning via RIM1 degradation [110] (Tables 3.4 and 3.5). A defect in FBXL20 leads to neurotransmission abnormalities, which could subsequently result in neurodegenerative phenotypes. *Fbxl20* knockout mice showed altered electrophysiological synaptic activity, which phenocopied RIM1 overexpression [110]. This observation is consistent with the notion that FBXL20 controls RIM1 degradation. Heterozygous *Fbxl20* mice showed differences in the contextual but not cued fear conditioning test, and also had an altered level of anxiety, suggesting a loss of FBXL20 specific function in the hippocampus [111]. Interestingly, mice overexpressing *Fbxl20* in the hippocampus did not show any significant difference in neurological/behavior testing (Table 3.3), indicating that although loss of function results in neurological alterations, an increase in FBXL20 activity is not sufficient to illicit any behavior phenotypes [111]. At the cellular level, an independent study found that *Fbxl20*-knockout mice exhibited long-term potentiation/depression (LTP/LTD) in CA3-CA1 synapses with smaller magnitudes after induction with low-frequency stimulation and LTPs with larger magnitudes after induction with tetanus stimulation (Table 3.2), suggesting that FBXL20 regulates metaplasticity [112]. Overall, FBXL20 mediated degradation appears to be important for hippocampal-dependent fear memory formation [111].

In a recent genome-wide association study (GWAS) to discover novel associations with estimated glomerular filtration rate (eGFR), a measure of kidney function, among the samples of European ancestry, a single-nucleotide polymorphism (SNP) in FBXL20 was identified [113]. However, further studies to understand the impact of this particular SNP on FBXL20 expression or function, and additional studies of FBXL20 in the kidney would need to be carried out to understand the role of FBXL20 in eGFR.

A role for FBXL20 in tumorigenesis was first identified in human colorectal adenocarcinoma, where mRNA levels of *FBXL20* were found to be upregulated in a large number of human colorectal adenocarcinoma tissues [114]. Furthermore, a direct role of FBXL20 was demonstrated by depleting *FBXL20* in cells line derived from these cancers, which lead to inhibition of cell proliferation, induction of apoptosis, and G1 cell cycle arrest, potentially through regulating Wnt signaling and caspase activation [114].

Finally, recent studies have identified that high miR-3151 expression was associated with shorter disease-free and overall survival in cytogenetically normal acute myeloid leukemia (CN-AML) patients [115]. In gene-expression profiling, high miR-3151 expressers showed downregulation of genes involved in transcriptional regulation, posttranslational modification, and cancer pathways [115]. *FBXL20* was one of two genes validated as direct miR-3151 targets, the other being the ubiquitin protease *USP40* [115]. In combination with the studies described above, these results indicate that FBXL20 may have an important role in tumorigenesis in a wide variety of cancer settings.

3.5   The Emerging Tumor Suppressor Role of FBXL21

Period determination in the mammalian circadian clock involves the turnover rate of the Cry and Per repressor proteins (Table 3.5). As discussed previously for FBXL3, ubiquitination of Cry1/2 mediates their degradation and thereby affecting the circadian clock [81, 82, 84]. Similar to FBXL3, a mutation in *FBXL21* (glycine to glutamate missense mutation) was identified (named *Past-time*) and characterized to have a short-period circadian [116]. Interestingly, a subsequent study found that FBXL21 also ubiquitinates Cry proteins. But rather than inducing degradation of Cry proteins, FBXL21 ubiquitination stabilizes Cry proteins by counteracting the effect of FBXL3 mediated Cry degradation. *Fbxl21* knockout mice exhibited normal periodicity but with compromised organization of daily activities [117] (Table 3.2). Consistent with counteracting FBXL3 regulation of circadian rhythm, the extremely long-period phenotype of *Fbxl3* mice was blocked in *Fbxl3/Fbxl21* double-knockout mice [117]. Unlike FBXL3, which resides in the nucleus, FBXL21 is found in the cytoplasm [116, 117]. It is striking to observe the opposing roles of FBXL21 on FBXL3 in the regulation of Cry stability and their ultimately combined effects at regulating the oscillation of the circadian clock. Notably, FBXL21 forms an SCF E3 ligase complex that slowly degrades Cry proteins in the cytoplasm but antagonizes the stronger E3 ligase activity of FBXL3 in the nucleus [116]. Therefore FBXL21 plays a dual role: protecting Cry proteins from FBXL3 degradation in the nucleus and promoting Cry protein degradation within the cytoplasm. Thus, the balance and cellular compartmentalization of competing E3 ligases targeting Cry proteins for degradation are involved in the circadian clock in mammals [116]. It will be of interest to tease out how these two FBXL factors are regulated themselves in the control of CRY ubiquitination and the timing of the circadian rhythm.

Unlike FBXL3, which is ubiquitously expressed, FBXL21 expression is tissue-specific [118]. Studies carried out in sheep, a diurnal ungulate, have also demonstrated that FBXL3 function is conserved and further revealed that its closest homologue, FBXL21, also binds to Cry1 which impaired its repressive action towards the transcriptional activators CLOCK/BMAL1 [118]. In sharp contrast with FBXL3, FBXL21 is highly expressed within the suprachiasmatic nuclei, site of the master clock, where it displays marked circadian oscillations apparently driven by members of the PAR-bZIP family [118]. Additionally, Dardente et al. identified and characterized splice variants of both FBXL3 and FBXL21 and functionally characterized novel splice-variants that might reduce Cry1 proteasomal degradation dependent on cellular context [118].

Similar to previous studies in humans with regard to FBXL3 and bipolar disorder, variants of the *FBXL21* gene have been associated with schizophrenia [119]. A significant association with schizophrenia was observed over two independent cohorts with three SNPs in *FBXL21* (rs31555, rs1859427, and rs1859427-rs6861170). Future studies on both FBXL3 and FBXL21 in circadian rhythm and cancer, as well as neurological disorders, is necessary to determine the potential of these factors as therapeutic targets.

To date, a potential role for FBXL21 in tumorigenesis and cancer progression has not yet been fully defined. With FBXL3 and FBXL21 sharing roles in the regulation of circadian rhythm (Tables 3.4 and 3.5), one might speculate that through this mechanism FBXL21 may play an important role in cancer development. However, extensive work is still necessary to elucidate whether FBXL21 functions as a tumor suppressor or oncoprotein given that similar to FBXL3, FBXL21 regulates Cry protein stability but the effect of FBXL21 on circadian rhythm is opposing that of FBXL3 [117]. Based on these ideas, FBXL21 might function as a tumor suppressor. However, identification of novel substrates of FBXL21 should allow for a more thorough analysis of a role for FBXL21 in cancer development.

References

1. Frescas D, Pagano M. Deregulated proteolysis by the F-box proteins SKP2 and beta-TrCP: tipping the scales of cancer. Nat Rev Cancer. 2008;8(6):438–49.
2. Jin J, et al. Systematic analysis and nomenclature of mammalian F-box proteins. Genes Dev. 2004;18(21):2573–80.
3. Chan CH, et al. Regulation of Skp2 expression and activity and its role in cancer progression. ScientificWorldJournal. 2010;10:1001–15.
4. Carrano AC, Pagano M. Role of the F-box protein Skp2 in adhesion-dependent cell cycle progression. J Cell Biol. 2001;153(7):1381–90.
5. Signoretti S, et al. Oncogenic role of the ubiquitin ligase subunit Skp2 in human breast cancer. J Clin Invest. 2002;110(5):633–41.

6. Waltregny D, et al. Androgen-driven prostate epithelial cell proliferation and differentiation in vivo involve the regulation of p27. Mol Endocrinol. 2001;15(5):765–82.
7. Lu L, Schulz H, Wolf DA. The F-box protein SKP2 mediates androgen control of p27 stability in LNCaP human prostate cancer cells. BMC Cell Biol. 2002;3:22.
8. Kullmann MK, et al. The p27-Skp2 axis mediates glucocorticoid-induced cell cycle arrest in T-lymphoma cells. Cell Cycle. 2013;12(16):2625–35.
9. Lim MS, et al. Expression of Skp2, a p27(Kip1) ubiquitin ligase, in malignant lymphoma: correlation with p27(Kip1) and proliferation index. Blood. 2002;100(8):2950–6.
10. Schuler S, et al. SKP2 confers resistance of pancreatic cancer cells towards TRAIL-induced apoptosis. Int J Oncol. 2011;38(1):219–25.
11. Hulit J, et al. p27Kip1 repression of ErbB2-induced mammary tumor growth in transgenic mice involves Skp2 and Wnt/beta-catenin signaling. Cancer Res. 2006;66(17):8529–41.
12. Fujita T, et al. Dissection of the APCCdh1-Skp2 cascade in breast cancer. Clin Cancer Res. 2008;14(7):1966–75.
13. Voduc D, et al. The combination of high cyclin E and Skp2 expression in breast cancer is associated with a poor prognosis and the basal phenotype. Hum Pathol. 2008;39(10):1431–7.
14. Liu J, et al. Cytoplasmic Skp2 expression is associated with p-Akt1 and predicts poor prognosis in human breast carcinomas. PLoS One. 2012;7(12):e52675.
15. Wei S, et al. Targeting the oncogenic E3 ligase Skp2 in prostate and breast cancer cells with a novel energy restriction-mimetic agent. PLoS One. 2012;7(10):e47298.
16. Zhao H, et al. Skp2 deletion unmasks a p27 safeguard that blocks tumorigenesis in the absence of pRb and p53 tumor suppressors. Cancer Cell. 2013;24(5):645–59.
17. Benevenuto-de-Andrade BA, et al. Immunohistochemical expression of Skp2 protein in oral nevi and melanoma. Med Oral Patol Oral Cir Bucal. 2013;18(3):e388–91.
18. Qu X, et al. A signal transduction pathway from TGF-beta1 to SKP2 via Akt1 and c-Myc and its correlation with progression in human melanoma. J Invest Dermatol. 2014;134:159–67.
19. Chen G, et al. Cytoplasmic Skp2 expression is increased in human melanoma and correlated with patient survival. PLoS One. 2011;6(2):e17578.
20. Xu HM, et al. Correlation of Skp2 overexpression to prognosis of patients with nasopharyngeal carcinoma from South China. Chin J Cancer. 2011;30(3):204–12.
21. Fang FM, et al. Effect of S-phase kinase-associated protein 2 expression on distant metastasis and survival in nasopharyngeal carcinoma patients. Int J Radiat Oncol Biol Phys. 2009;73(1):202–7.
22. Masuda TA, et al. Clinical and biological significance of S-phase kinase-associated protein 2 (Skp2) gene expression in gastric carcinoma: modulation of malignant phenotype by Skp2 overexpression, possibly via p27 proteolysis. Cancer Res. 2002;62(13):3819–25.
23. Lu M, et al. The expression and prognosis of FOXO3a and Skp2 in human hepatocellular carcinoma. Pathol Oncol Res. 2009;15(4):679–87.
24. Lin HK, et al. Skp2 targeting suppresses tumorigenesis by Arf-p53-independent cellular senescence. Nature. 2010;464(7287):374–9.
25. Radke S, Pirkmaier A, Germain D. Differential expression of the F-box proteins Skp2 and Skp2B in breast cancer. Oncogene. 2005;24(21):3448–58.
26. Nakayama KI, Hatakeyama S, Nakayama K. Regulation of the cell cycle at the G1-S transition by proteolysis of cyclin E and p27Kip1. Biochem Biophys Res Commun. 2001;282(4):853–60.
27. Suzuki S, et al. The amelioration of renal damage in Skp2-deficient mice canceled by p27 Kip1 deficiency in Skp2−/− p27−/− mice. PLoS One. 2012;7(4):e36249.
28. Egozi D, et al. Regulation of the cell cycle inhibitor p27 and its ubiquitin ligase Skp2 in differentiation of human embryonic stem cells. FASEB J. 2007;21(11):2807–17.
29. Dombrowski C, et al. FGFR1 signaling stimulates proliferation of human mesenchymal stem cells by inhibiting the cyclin-dependent kinase inhibitors P21 and P27. Stem Cells. 2013;31:2724–36.
30. Kitagawa K, Kotake Y, Kitagawa M. Ubiquitin-mediated control of oncogene and tumor suppressor gene products. Cancer Sci. 2009;100(8):1374–81.

31. Nakayama K, et al. Targeted disruption of Skp2 results in accumulation of cyclin E and p27(Kip1), polyploidy and centrosome overduplication. EMBO J. 2000;19(9):2069–81.
32. Nakayama K, et al. Skp2-mediated degradation of p27 regulates progression into mitosis. Dev Cell. 2004;6(5):661–72.
33. Carrano AC, et al. SKP2 is required for ubiquitin-mediated degradation of the CDK inhibitor p27. Nat Cell Biol. 1999;1(4):193–9.
34. Tsvetkov LM, et al. p27(Kip1) ubiquitination and degradation is regulated by the SCF(Skp2) complex through phosphorylated Thr187 in p27. Curr Biol. 1999;9(12):661–4.
35. Spruck C, et al. A CDK-independent function of mammalian Cks1: targeting of SCF(Skp2) to the CDK inhibitor p27Kip1. Mol Cell. 2001;7(3):639–50.
36. Ganoth D, et al. The cell-cycle regulatory protein Cks1 is required for SCF(Skp2)-mediated ubiquitinylation of p27. Nat Cell Biol. 2001;3(3):321–4.
37. Yu ZK, Gervais JL, Zhang H. Human CUL-1 associates with the SKP1/SKP2 complex and regulates p21(CIP1/WAF1) and cyclin D proteins. Proc Natl Acad Sci U S A. 1998; 95(19):11324–9.
38. Kamura T, et al. Degradation of p57Kip2 mediated by SCFSkp2-dependent ubiquitylation. Proc Natl Acad Sci U S A. 2003;100(18):10231–6.
39. Hiramatsu Y, et al. Degradation of Tob1 mediated by SCFSkp2-dependent ubiquitination. Cancer Res. 2006;66(17):8477–83.
40. Song MS, et al. Skp2 regulates the antiproliferative function of the tumor suppressor RASSF1A via ubiquitin-mediated degradation at the G1-S transition. Oncogene. 2008; 27(22):3176–85.
41. Bhattacharya S, et al. SKP2 associates with p130 and accelerates p130 ubiquitylation and degradation in human cells. Oncogene. 2003;22(16):2443–51.
42. Huang H, et al. Skp2 inhibits FOXO1 in tumor suppression through ubiquitin-mediated degradation. Proc Natl Acad Sci U S A. 2005;102(5):1649–54.
43. Wang H, et al. A comparison between Skp2 and FOXO1 for their cytoplasmic localization by Akt1. Cell Cycle. 2010;9(5):1021–2.
44. Xu F, et al. The oncoprotein HBXIP up-regulates Skp2 via activating transcription factor E2F1 to promote proliferation of breast cancer cells. Cancer Lett. 2013;333(1):124–32.
45. Zhang L, Wang C. F-box protein Skp2: a novel transcriptional target of E2F. Oncogene. 2006;25(18):2615–27.
46. Fukushima H, et al. SCF(Fbw7) modulates the NFkB signaling pathway by targeting NFkB2 for ubiquitination and destruction. Cell Rep. 2012;1(5):434–43.
47. Appleman LJ, et al. CD28 costimulation mediates transcription of SKP2 and CKS1, the substrate recognition components of SCFSkp2 ubiquitin ligase that leads p27kip1 to degradation. Cell Cycle. 2006;5(18):2123–9.
48. Sarmento LM, et al. Notch1 modulates timing of G1-S progression by inducing SKP2 transcription and p27 Kip1 degradation. J Exp Med. 2005;202(1):157–68.
49. Imaki H, et al. Cell cycle-dependent regulation of the Skp2 promoter by GA-binding protein. Cancer Res. 2003;63(15):4607–13.
50. Huang H, Zhao W, Yang D. Stat3 induces oncogenic Skp2 expression in human cervical carcinoma cells. Biochem Biophys Res Commun. 2012;418(1):186–90.
51. Wei Z, et al. STAT3 interacts with Skp2/p27/p21 pathway to regulate the motility and invasion of gastric cancer cells. Cell Signal. 2013;25(4):931–8.
52. Park TJ, et al. Skp2 enhances polyubiquitination and degradation of TIS21/BTG2/PC3, tumor suppressor protein, at the downstream of FoxM1. Exp Cell Res. 2009;315(18): 3152–62.
53. Zuo T, et al. FOXP3 is a novel transcriptional repressor for the breast cancer oncogene SKP2. J Clin Invest. 2007;117(12):3765–73.
54. Zheng J, et al. Heterozygous genetic variations of FOXP3 in Xp11.23 elevate breast cancer risk in Chinese population via skewed X-chromosome inactivation. Hum Mutat. 2013;34(4):619–28.

55. Bashir T, et al. Control of the SCF(Skp2-Cks1) ubiquitin ligase by the APC/C(Cdh1) ubiquitin ligase. Nature. 2004;428(6979):190–3.
56. Kurland JF, Tansey WP. Crashing waves of destruction: the cell cycle and APC(Cdh1) regulation of SCF(Skp2). Cancer Cell. 2004;5(4):305–6.
57. Wei W, et al. Degradation of the SCF component Skp2 in cell-cycle phase G1 by the anaphase-promoting complex. Nature. 2004;428(6979):194–8.
58. Fang Y, et al. MicroRNA-7 inhibits tumor growth and metastasis by targeting the phosphoinositide 3-kinase/Akt pathway in hepatocellular carcinoma. Hepatology. 2012;55(6):1852–62.
59. Sanchez N, et al. MiR-7 triggers cell cycle arrest at the G1/S transition by targeting multiple genes including Skp2 and Psme3. PLoS One. 2013;8(6):e65671.
60. Jackson SJ, et al. Rapid and widespread suppression of self-renewal by microRNA-203 during epidermal differentiation. Development. 2013;140(9):1882–91.
61. Wang H, et al. An AR-Skp2 pathway for proliferation of androgen-dependent prostate-cancer cells. J Cell Sci. 2008;121(Pt 15):2578–87.
62. Jiang J, et al. Androgens repress expression of the F-box protein Skp2 via p107 dependent and independent mechanisms in LNCaP prostate cancer cells. Prostate. 2012;72(2):225–32.
63. Pernicova Z, et al. Androgen depletion induces senescence in prostate cancer cells through down-regulation of Skp2. Neoplasia. 2011;13(6):526–36.
64. Kratzat S, et al. Cks1 is required for tumor cell proliferation but not sufficient to induce hematopoietic malignancies. PLoS One. 2012;7(5):e37433.
65. Agarwal A, et al. Absence of SKP2 expression attenuates BCR-ABL-induced myeloproliferative disease. Blood. 2008;112(5):1960–70.
66. Minamishima YA, Nakayama K. Recovery of liver mass without proliferation of hepatocytes after partial hepatectomy in Skp2-deficient mice. Cancer Res. 2002;62(4):995–9.
67. Chander H, et al. Skp2B attenuates p53 function by inhibiting prohibitin. EMBO Rep. 2010;11(3):220–5.
68. Shim EH, et al. Expression of the F-box protein SKP2 induces hyperplasia, dysplasia, and low-grade carcinoma in the mouse prostate. Cancer Res. 2003;63(7):1583–8.
69. Sistrunk C, et al. Skp2 deficiency inhibits chemical skin tumorigenesis independent of p27(Kip1) accumulation. Am J Pathol. 2013;182(5):1854–64.
70. Koepp DM, et al. Phosphorylation-dependent ubiquitination of cyclin E by the SCFFbw7 ubiquitin ligase. Science. 2001;294(5540):173–7.
71. Wang H, et al. Skp2 is required for survival of aberrantly proliferating Rb1-deficient cells and for tumorigenesis in Rb1+/− mice. Nat Genet. 2010;42(1):83–8.
72. Pagano M. Control of DNA synthesis and mitosis by the Skp2-p27-Cdk1/2 axis. Mol Cell. 2004;14(4):414–6.
73. Rico-Bautista E, Wolf DA. Skipping cancer: small molecule inhibitors of SKP2-mediated p27 degradation. Chem Biol. 2012;19(12):1497–8.
74. Chen Q, et al. Targeting the p27 E3 ligase SCF(Skp2) results in p27- and Skp2-mediated cell-cycle arrest and activation of autophagy. Blood. 2008;111(9):4690–9.
75. Rico-Bautista E, et al. Chemical genetics approach to restoring p27Kip1 reveals novel compounds with antiproliferative activity in prostate cancer cells. BMC Biol. 2010;8:153.
76. Wu L, et al. Specific small molecule inhibitors of Skp2-mediated p27 degradation. Chem Biol. 2012;19(12):1515–24.
77. Ju Y, et al. Glucosamine, a naturally occurring amino monosaccharide, inhibits A549 and H446 cell proliferation by blocking G1/S transition. Mol Med Rep. 2013;8(3):794–8.
78. Koo KH, et al. Salinomycin induces cell death via inactivation of Stat3 and downregulation of Skp2. Cell Death Dis. 2013;4:e693.
79. Chan CH, et al. Pharmacological inactivation of Skp2 SCF ubiquitin ligase restricts cancer stem cell traits and cancer progression. Cell. 2013;154(3):556–68.
80. Chan CH, et al. The Skp2-SCF E3 ligase regulates Akt ubiquitination, glycolysis, herceptin sensitivity, and tumorigenesis. Cell. 2012;149(5):1098–111.

81. Godinho SI, et al. The after-hours mutant reveals a role for Fbxl3 in determining mammalian circadian period. Science. 2007;316(5826):897–900.
82. Siepka SM, et al. Circadian mutant overtime reveals F-box protein FBXL3 regulation of cryptochrome and period gene expression. Cell. 2007;129(5):1011–23.
83. Vashisht AA, et al. Control of iron homeostasis by an iron-regulated ubiquitin ligase. Science. 2009;326(5953):718–21.
84. Busino L, et al. SCFFbxl3 controls the oscillation of the circadian clock by directing the degradation of cryptochrome proteins. Science. 2007;316(5826):900–4.
85. Xing W, et al. SCF(FBXL3) ubiquitin ligase targets cryptochromes at their cofactor pocket. Nature. 2013;496(7443):64–8.
86. Tong X, et al. USP2a protein deubiquitinates and stabilizes the circadian protein CRY1 in response to inflammatory signals. J Biol Chem. 2012;287(30):25280–91.
87. Shi G, et al. Dual roles of FBXL3 in the mammalian circadian feedback loops are important for period determination and robustness of the clock. Proc Natl Acad Sci U S A. 2013;110(12):4750–5.
88. Ahn YM, et al. Chronotype distribution in bipolar I disorder and schizophrenia in a Korean sample. Bipolar Disord. 2008;10(2):271–5.
89. Mansour HA, et al. Circadian phase variation in bipolar I disorder. Chronobiol Int. 2005;22(3):571–84.
90. McClung CA. Circadian genes, rhythms and the biology of mood disorders. Pharmacol Ther. 2007;114(2):222–32.
91. Wood J, et al. Replicable differences in preferred circadian phase between bipolar disorder patients and control individuals. Psychiatry Res. 2009;166(2–3):201–9.
92. Keers R, et al. Reduced anxiety and depression-like behaviours in the circadian period mutant mouse afterhours. PLoS One. 2012;7(6):e38263.
93. Williams DS, et al. Nonsense mediated decay resistant mutations are a source of expressed mutant proteins in colon cancer cell lines with microsatellite instability. PLoS One. 2010;5(12):e16012.
94. Luo Y, et al. Deregulated expression of cry1 and cry2 in human gliomas. Asian Pac J Cancer Prev. 2012;13(11):5725–8.
95. Savvidis C, Koutsilieris M. Circadian rhythm disruption in cancer biology. Mol Med. 2012;18:1249–60.
96. Fu L, et al. The circadian gene Period2 plays an important role in tumor suppression and DNA damage response in vivo. Cell. 2002;111(1):41–50.
97. Gery S, et al. The circadian gene per1 plays an important role in cell growth and DNA damage control in human cancer cells. Mol Cell. 2006;22(3):375–82.
98. Schmutz I, et al. The mammalian clock component PERIOD2 coordinates circadian output by interaction with nuclear receptors. Genes Dev. 2010;24(4):345–57.
99. Filipski E, et al. Effects of chronic jet lag on tumor progression in mice. Cancer Res. 2004;64(21):7879–85.
100. Kelleher FC, Rao A, Maguire A. Circadian molecular clocks and cancer. Cancer Lett. 2014;342(1):9–18.
101. Ozturk N, et al. Loss of cryptochrome reduces cancer risk in p53 mutant mice. Proc Natl Acad Sci U S A. 2009;106(8):2841–6.
102. Unsal-Kacmaz K, et al. Coupling of human circadian and cell cycles by the timeless protein. Mol Cell Biol. 2005;25(8):3109–16.
103. Tsukada Y, et al. Histone demethylation by a family of JmjC domain-containing proteins. Nature. 2006;439(7078):811–6.
104. Wu X, Johansen JV, Helin K. Fbxl10/Kdm2b recruits polycomb repressive complex 1 to CpG islands and regulates H2A ubiquitylation. Mol Cell. 2013;49(6):1134–46.
105. Gearhart MD, et al. Polycomb group and SCF ubiquitin ligases are found in a novel BCOR complex that is recruited to BCL6 targets. Mol Cell Biol. 2006;26(18):6880–9.

106. Sanchez C, et al. Proteomics analysis of Ring1B/Rnf2 interactors identifies a novel complex with the Fbxl10/Jhdm1B histone demethylase and the Bcl6 interacting corepressor. Mol Cell Proteomics. 2007;6(5):820–34.
107. Huynh KD, et al. BCoR, a novel corepressor involved in BCL-6 repression. Genes Dev. 2000;14(14):1810–23.
108. Frescas D, et al. JHDM1B/FBXL10 is a nucleolar protein that represses transcription of ribosomal RNA genes. Nature. 2007;450(7167):309–13.
109. Tzatsos A, et al. KDM2B promotes pancreatic cancer via Polycomb-dependent and -independent transcriptional programs. J Clin Invest. 2013;123(2):727–39.
110. Yao I, et al. SCRAPPER-dependent ubiquitination of active zone protein RIM1 regulates synaptic vesicle release. Cell. 2007;130(5):943–57.
111. Yao I, et al. Synaptic E3 ligase SCRAPPER in contextual fear conditioning: extensive behavioral phenotyping of Scrapper heterozygote and overexpressing mutant mice. PLoS One. 2011;6(2):e17317.
112. Takagi H, et al. SCRAPPER regulates the thresholds of long-term potentiation/depression, the bidirectional synaptic plasticity in hippocampal CA3-CA1 synapses. Neural Plast. 2012;2012:352829.
113. Chasman DI, et al. Integration of genome-wide association studies with biological knowledge identifies six novel genes related to kidney function. Hum Mol Genet. 2012;21(24):5329–43.
114. Zhu J, et al. Role of FBXL20 in human colorectal adenocarcinoma. Oncol Rep. 2012;28(6):2290–8.
115. Eisfeld AK, et al. miR-3151 interplays with its host gene BAALC and independently affects outcome of patients with cytogenetically normal acute myeloid leukemia. Blood. 2012;120(2):249–58.
116. Yoo SH, et al. Competing E3 ubiquitin ligases govern circadian periodicity by degradation of CRY in nucleus and cytoplasm. Cell. 2013;152(5):1091–105.
117. Hirano A, et al. FBXL21 regulates oscillation of the circadian clock through ubiquitination and stabilization of cryptochromes. Cell. 2013;152(5):1106–18.
118. Dardente H, et al. Implication of the F-Box Protein FBXL21 in circadian pacemaker function in mammals. PLoS One. 2008;3(10):e3530.
119. Chen X, et al. FBXL21 association with schizophrenia in Irish family and case-control samples. Am J Med Genet B Neuropsychiatr Genet. 2008;147B(7):1231–7.
120. Kossatz U, et al. Skp2-dependent degradation of p27kip1 is essential for cell cycle progression. Genes Dev. 2004;18(21):2602–7.
121. Latres E, et al. Role of the F-box protein Skp2 in lymphomagenesis. Proc Natl Acad Sci U S A. 2001;98(5):2515–20.
122. Gstaiger M, et al. Skp2 is oncogenic and overexpressed in human cancers. Proc Natl Acad Sci U S A. 2001;98(9):5043–8.
123. Yokoi S, et al. Amplification and overexpression of SKP2 are associated with metastasis of non-small-cell lung cancers to lymph nodes. Am J Pathol. 2004;165(1):175–80.
124. Chiaur DS, et al. Five human genes encoding F-box proteins: chromosome mapping and analysis in human tumors. Cytogenet Cell Genet. 2000;88(3–4):255–8.
125. Fukuda T, et al. Fbxl10/Kdm2b deficiency accelerates neural progenitor cell death and leads to exencephaly. Mol Cell Neurosci. 2011;46(3):614–24.
126. Suzuki S, et al. Renal damage in obstructive nephropathy is decreased in Skp2-deficient mice. Am J Pathol. 2007;171(2):473–83.
127. Fotovati A, et al. Impaired ovarian development and reduced fertility in female mice deficient in Skp2. J Anat. 2011;218(6):668–77.
128. Wu YJ, et al. S-phase kinase-associated protein-2 (Skp2) promotes vascular smooth muscle cell proliferation and neointima formation in vivo. J Vasc Surg. 2009;50(5):1135–42.
129. Zhong L, et al. Essential role of Skp2-mediated p27 degradation in growth and adaptive expansion of pancreatic beta cells. J Clin Invest. 2007;117(10):2869–76.

130. Sakai T, et al. Skp2 controls adipocyte proliferation during the development of obesity. J Biol Chem. 2007;282(3):2038–46.
131. Fotovati A, Nakayama K, Nakayama KI. Impaired germ cell development due to compromised cell cycle progression in Skp2-deficient mice. Cell Div. 2006;1:4.
132. Wang J, et al. The role of Skp2 in hematopoietic stem cell quiescence, pool size, and self-renewal. Blood. 2011;118(20):5429–38.
133. Umanskaya K, et al. Skp2B stimulates mammary gland development by inhibiting REA, the repressor of the estrogen receptor. Mol Cell Biol. 2007;27(21):7615–22.
134. Sistrunk C, et al. Skp2 is necessary for Myc-induced keratinocyte proliferation but dispensable for Myc oncogenic activity in the oral epithelium. Am J Pathol. 2011;178(6):2470–7.
135. Watanabe T, et al. FBL2 regulates amyloid precursor protein (APP) metabolism by promoting ubiquitination-dependent APP degradation and inhibition of APP endocytosis. J Neurosci. 2012;32(10):3352–65.
136. Nie L, et al. Notch-induced E2A ubiquitination and degradation are controlled by MAP kinase activities. EMBO J. 2003;22(21):5780–92.
137. Vaites LP, et al. The Fbx4 tumor suppressor regulates cyclin D1 accumulation and prevents neoplastic transformation. Mol Cell Biol. 2011;31(22):4513–23.
138. Hardisty-Hughes RE, et al. A mutation in the F-box gene, Fbxo11, causes otitis media in the Jeff mouse. Hum Mol Genet. 2006;15(22):3273–9.
139. Tetzlaff MT, et al. Cyclin F disruption compromises placental development and affects normal cell cycle execution. Mol Cell Biol. 2004;24(6):2487–98.
140. Lee TH, et al. The F-box protein FBX4 targets PIN2/TRF1 for ubiquitin-mediated degradation and regulates telomere maintenance. J Biol Chem. 2006;281(2):759–68.
141. Ruiz JC, et al. F-box and leucine-rich repeat protein 5 (FBXL5) is required for maintenance of cellular and systemic iron homeostasis. J Biol Chem. 2013;288(1):552–60.
142. Nelson RF, et al. Selective cochlear degeneration in mice lacking the F-box protein, Fbx2, a glycoprotein-specific ubiquitin ligase subunit. J Neurosci. 2007;27(19):5163–71.
143. Tokuzawa Y, et al. Fbx15 is a novel target of Oct3/4 but is dispensable for embryonic stem cell self-renewal and mouse development. Mol Cell Biol. 2003;23(8):2699–708.
144. Saiga T, et al. Fbxo45 forms a novel ubiquitin ligase complex and is required for neuronal development. Mol Cell Biol. 2009;29(13):3529–43.
145. D'Angiolella V, et al. Cyclin F-mediated degradation of ribonucleotide reductase M2 controls genome integrity and DNA repair. Cell. 2012;149(5):1023–34.
146. Yoshida Y, et al. E3 ubiquitin ligase that recognizes sugar chains. Nature. 2002;418(6896):438–42.
147. Murai-Takebe R, et al. Ubiquitination-mediated regulation of biosynthesis of the adhesion receptor SHPS-1 in response to endoplasmic reticulum stress. J Biol Chem. 2004;279(12):11616–25.
148. Kato A, et al. Activity-dependent NMDA receptor degradation mediated by retrotranslocation and ubiquitination. Proc Natl Acad Sci U S A. 2005;102(15):5600–5.
149. Henzl MT, et al. The cochlear F-box protein OCP1 associates with OCP2 and connexin 26. Hear Res. 2004;191(1–2):101–9.
150. Liu B, et al. Proteomic identification of common SCF ubiquitin ligase FBXO6-interacting glycoproteins in three kinds of cells. J Proteome Res. 2012;11(3):1773–81.
151. Shima Y, et al. PML activates transcription by protecting HIPK2 and p300 from SCFFbx3-mediated degradation. Mol Cell Biol. 2008;28(23):7126–38.
152. Jia L, Sun Y. F-box proteins FBXO31 and FBX4 in regulation of cyclin D1 degradation upon DNA damage. Pigment Cell Melanoma Res. 2009;22(5):518–9.
153. Lin DI, et al. Phosphorylation-dependent ubiquitination of cyclin D1 by the SCF(FBX4-alphaB crystallin) complex. Mol Cell. 2006;24(3):355–66.
154. Santra MK, Wajapeyee N, Green MR. F-box protein FBXO31 mediates cyclin D1 degradation to induce G1 arrest after DNA damage. Nature. 2009;459(7247):722–5.
155. Yoshida Y, et al. Fbs2 is a new member of the E3 ubiquitin ligase family that recognizes sugar chains. J Biol Chem. 2003;278(44):43877–84.

156. Zhang YW, et al. The F box protein Fbx6 regulates Chk1 stability and cellular sensitivity to replication stress. Mol Cell. 2009;35(4):442–53.
157. Merry C, et al. Targeting the checkpoint kinase Chk1 in cancer therapy. Cell Cycle. 2010;9(2):279–83.
158. Agrawal N, et al. Exome sequencing of head and neck squamous cell carcinoma reveals inactivating mutations in NOTCH1. Science. 2011;333(6046):1154–7.
159. Arabi A, et al. Proteomic screen reveals Fbw7 as a modulator of the NF-kappaB pathway. Nat Commun. 2012;3:976.
160. Balamurugan K, et al. FBXW7alpha attenuates inflammatory signalling by downregulating C/EBPdelta and its target gene Tlr4. Nat Commun. 2013;4:1662.
161. Seki A, et al. Plk1- and beta-TrCP-dependent degradation of Bora controls mitotic progression. J Cell Biol. 2008;181(1):65–78.
162. Kuchay S, et al. FBXL2- and PTPL1-mediated degradation of p110-free p85beta regulatory subunit controls the PI(3)K signalling cascade. Nat Cell Biol. 2013;15(5):472–80.
163. Chen BB, et al. F-box protein FBXL2 targets cyclin D2 for ubiquitination and degradation to inhibit leukemic cell proliferation. Blood. 2012;119(13):3132–41.
164. Chen BB, et al. FBXL2 is a ubiquitin E3 ligase subunit that triggers mitotic arrest. Cell Cycle. 2011;10(20):3487–94.
165. Chen BB, et al. F-box protein FBXL2 exerts human lung tumor suppressor-like activity by ubiquitin-mediated degradation of cyclin D3 resulting in cell cycle arrest. Oncogene. 2012; 31(20):2566–79.
166. Maywood ES, et al. Genetic and molecular analysis of the central and peripheral circadian clockwork of mice. Cold Spring Harb Symp Quant Biol. 2007;72:85–94.
167. Van Rechem C, et al. The SKP1-Cul1-F-box and leucine-rich repeat protein 4 (SCF-FbxL4) ubiquitin ligase regulates lysine demethylase 4A (KDM4A)/Jumonji domain-containing 2A (JMJD2A) protein. J Biol Chem. 2011;286(35):30462–70.
168. Salahudeen AA, et al. An E3 ligase possessing an iron-responsive hemerythrin domain is a regulator of iron homeostasis. Science. 2009;326(5953):722–6.
169. Coon TA, et al. Novel E3 ligase component FBXL7 ubiquitinates and degrades Aurora A, causing mitotic arrest. Cell Cycle. 2012;11(4):721–9.
170. Postow L, Funabiki H. An SCF complex containing Fbxl12 mediates DNA damage-induced Ku80 ubiquitylation. Cell Cycle. 2013;12(4):587–95.
171. Mallampalli RK, et al. Fbxl12 triggers G1 arrest by mediating degradation of calmodulin kinase I. Cell Signal. 2013;25(10):2047–59.
172. Vinas-Castells R, et al. The hypoxia-controlled FBXL14 ubiquitin ligase targets SNAIL1 for proteasome degradation. J Biol Chem. 2010;285(6):3794–805.
173. Vernon AE, LaBonne C. Slug stability is dynamically regulated during neural crest development by the F-box protein Ppa. Development. 2006;133(17):3359–70.
174. Zheng H, et al. Essential role of Fbxl14 ubiquitin ligase in regulation of vertebrate axis formation through modulating Mkp3 level. Cell Res. 2012;22(5):936–40.
175. Lander R, Nordin K, LaBonne C. The F-box protein Ppa is a common regulator of core EMT factors Twist, Snail, Slug, and Sip1. J Cell Biol. 2011;194(1):17–25.
176. Cui Y, et al. SCFFBXL(1)(5) regulates BMP signalling by directing the degradation of HECT-type ubiquitin ligase Smurf1. EMBO J. 2011;30(13):2675–89.
177. Zhao J, et al. F-box protein FBXL19-mediated ubiquitination and degradation of the receptor for IL-33 limits pulmonary inflammation. Nat Immunol. 2012;13(7):651–8.
178. Zhao J, et al. SCF E3 ligase F-box protein complex SCFFBXL19 regulates cell migration by mediating Rac1 ubiquitination and degradation. FASEB J. 2013;27(7):2611–9.
179. Wei J, et al. A new mechanism of RhoA ubiquitination and degradation: roles of SCF E3 ligase and Erk2. Biochim Biophys Acta. 2013;1833(12):2757–64.
180. Spaich S, et al. F-box and leucine-rich repeat protein 22 is a cardiac-enriched F-box protein that regulates sarcomeric protein turnover and is essential for maintenance of contractile function in vivo. Circ Res. 2012;111(12):1504–16.

Chapter 4
The Role of FBXO Subfamily
of F-box Proteins in Tumorigenesis

Jianping Guo, Brian J. North, Adriana E. Tron, Hiroyuki Inuzuka,
and Wenyi Wei

Abstract Within the 69 identified putative F-box proteins in the human genome, besides the FBXW subfamily (Chap. 2) and the FBXL subfamily (Chap. 3), the remaining 36 F-box proteins were designated as F-box only (FBXO) proteins. It is noteworthy that the FBXO subfamily proteins have been identified to contain the F-box motif in its N-terminus and various types of functional domains in its C-terminus that mediate substrate binding. Unlike FBXW proteins with the WD40 repeats motif and FBXL proteins with the LRR motif, most of the functional domains in FBXO proteins have yet to be fully characterized. Compared to the FBXW and FBXL subfamilies, the FBXO subfamily is more diversified with 21 functional homology domains identified within the FBXO subfamily (Fig. 4.1).

In this chapter, we focus our discussion on the recent genetic, pathological as well as the biochemical evidence suggesting possible tumor suppressor or oncogenic roles for FBXO subfamily members (Tables 4.1 and 4.3). As stated in previous chapters, given the fact that physiological evidence (mouse modeling results) is considered as the strongest supportive data to implicate any given F-box protein in tumorigenesis (Table 4.2), we limit our discussion to those FBXO members with available mouse genetic models.

Keywords F-box • SCF • FBXO subfamily • FBXO1 • Cyclin F • FBXO4 • FBXO7 • FBXO32 • Cullin 1 • Cyclin D • Muscle • Tumor suppressor • Oncoprotein • Mouse model • Physiological function

J. Guo • B.J. North • A.E. Tron • H. Inuzuka (✉) • W. Wei (✉)
Department of Pathology, Beth Israel Deaconess Medical Center,
Harvard Medical School, Boston, MA 02215, USA
e-mail: hinuzuka@bidmc.harvard.edu; wwei2@bidmc.harvard.edu

H. Inuzuka and W. Wei, *SCF and APC E3 Ubiquitin Ligases in Tumorigenesis*,
SpringerBriefs in Cancer Research, DOI 10.1007/978-3-319-05026-3_4,
© The Author(s) 2014

4.1   Emerging Tumor Suppressor Role of FBXO1

FBXO1 (also known as FBX1 or Cyclin F) was initially identified as a novel mammalian cyclin capable of suppressing the temperature sensitivity of a *Saccharomyces cerevisiae Cdc4* mutant [1]. FBXO1, localizes to both the centrosome and nucleus, shares the greatest amino acid sequence similarity with Cyclin A, and oscillates during the cell cycle, accumulating in S phase, peaking in G2 phase, and disappearing as cells enter mitosis [2]. However, unlike other cyclins, which are tightly controlled and degraded by SCF ubiquitin ligase complexes, FBXO1 degradation is independent of ubiquitination and proteasome-mediated pathways and the molecular mechanism governing its degradation remains unclear [2]. With tightly regulated expression and high conservation from amphibians to mammals suggest that FBXO1 is a critical regulator of cell cycle progression, but unlike a majority of traditional cyclins, FBXO1 does not bind or activate any *cyclin*-dependent *k*inases (Cdks), and its function in cell cycle regulation is largely unclear [2]. One proposed function for FBXO1 is through its ability to bind Cyclin B and transport it into the nucleus, which is the first example of a direct cyclin-cyclin regulatory mechanism, and elucidates a novel regulatory mechanism governing Cyclin B/Cdk1 localization and function during mitosis [3] (Fig. 4.1 and Table 4.1).

In addition to harboring a cyclin domain, a more important functional domain within FBXO1 is its F-box motif that is the founding member of the large family of proteins containing the F-box motif in all eukaryotes [4]. Similar to other F-box proteins, FBXO1 binds to Skp1, and at the same time, Skp1 recruits Cullin 1 and RBX1 to assemble a functional SCF complex. RBX1 recruits the E2 for target substrate ubiquitination, which are recruited to the complex by interaction with FBXO1. Thus far, three proteins (CP110, NuSAP1, and RRM2) have been identified as substrates of FBXO1-mediated proteasome degradation, and targeted by FBXO1 during the G2 phase of the cell cycle [5–7] (Table 4.3). FBXO1 functions to control genomic stability through inhibiting centrosome reduplication by targeting CP110 for degradation, and depletion of FBXO1 induces centrosomal and mitotic abnormalities [5]. Furthermore, FBXO1 maintained the homeostasis of dNTP pools by degrading Cdk-mediated-phosphorylated RRM2 (the *r*ibonucleotide *r*eductase family *m*ember 2), which regulates dNTP balance and is necessary for both replicative and repair based DNA synthesis [7]. Moreover, in response to DNA damage, FBXO1 is downregulated in an ATR-dependent manner to allow accumulation of RRM2, and depletion of FBXO1 leads to a delay in DNA repair and sensitizes cells to DNA damage [7]. With respect to cancer treatment, FBXO1 sensitized cells to microtubule-based chemotherapeutics by targeting NuSAP1, a cell-cycle-regulated microtubule-binding protein with roles in chromosome congression and segregation, and mitotic spindle organization [6]. Taken together, these studies highlight the direct role for FBXO1 in controlling genome stability through ubiquitin-mediated proteolysis and implications for cancer development and therapy.

Subsequent studies using mouse knockout models showed that mice carrying a homozygous deletion were embryonic lethal at day E10.5 exhibiting developmental anomalies, whereas heterozygous mice appeared normal (Table 4.2). MEFs lacking

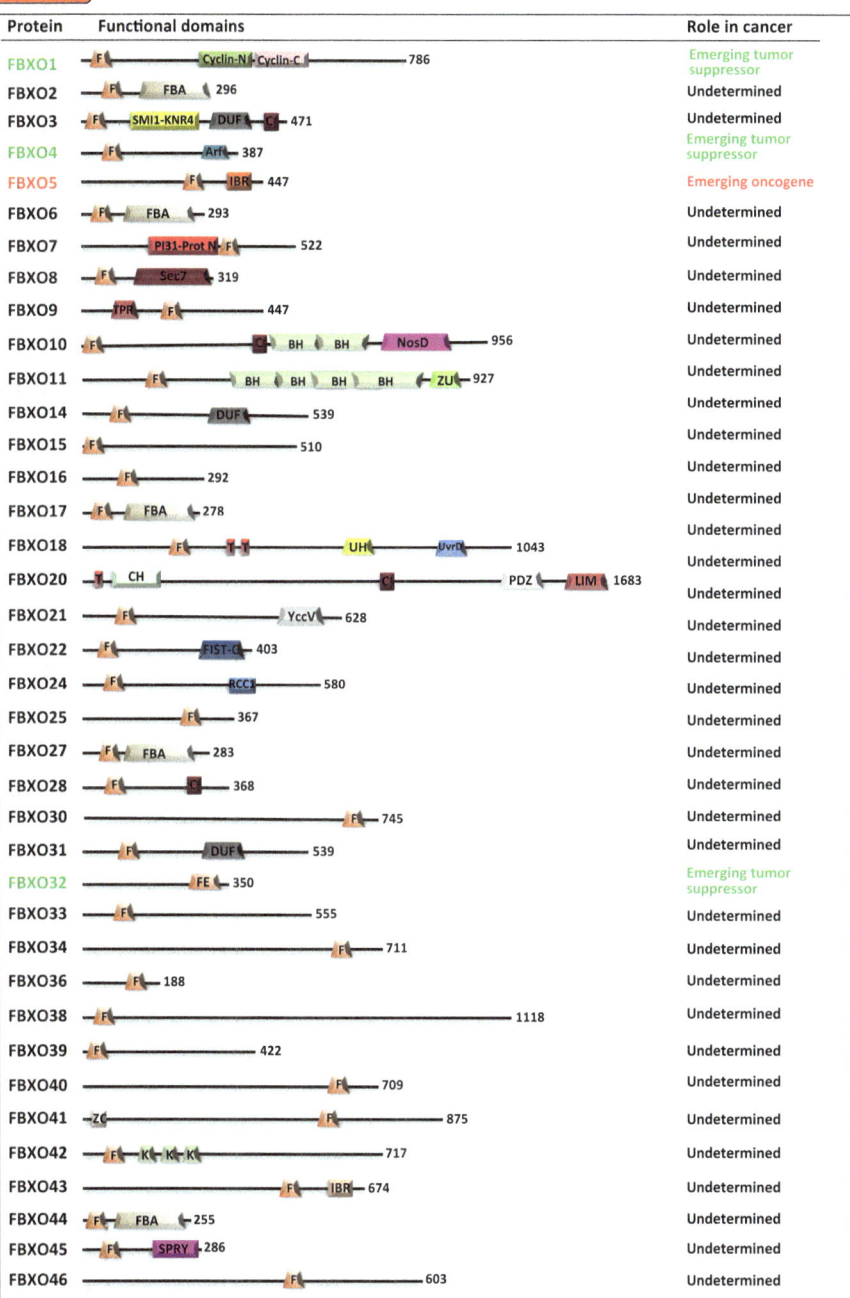

Fig. 4.1 A schematic illustration of functional domains of all known FBXO proteins and their possible roles in cancer. *F* F-box motif, *Cyclin-N* cyclin domain, *Cyclin-C* cyclin domain, *FBA* F-box associated domain, *PI31-Prot* PI31 proteasome regulator N-terminal domain, *SPRY* spla and the ryanodine receptor domain, *IBR* in between ring fingers domain, *SMI1-KNR* SMI1/KNR4 family (SUKH-1), *Sec7* GEF for the PF00025 family, *TPR* tetratricopeptide-like repeats, *BH* beta helix, *NosD* periplasmic copper-binding protein, *Zu* putative zinc finger in N-recognin, *UvrD* REP helicase N-terminal domain, *UH* UvrD helicase, *CH* calponin homology domain, *LIM* LIM domain, *YccV* YccV-like domain, *FIST-C* sensory domain, *Nop14* Nop14-like family domain, *Nu* Nucleoporin Nup120/160-like domain, *DUF* protein of unknown function, *K* Kelch motif

Table 4.1 Summary of the roles of the FBXO class of F-box proteins in cancer

Potential role in cancer (tumor suppressor or oncogene)	F-box protein	Gene symbol	Physiological evidence (mouse models)		Pathological evidence in cancer		Biochemical evidence (major substrates)
			Knockout	Transgenic	Overexpression	Mutation/deletion	
Emerging tumor suppressor	FBXO1	CCNF	Yes [8]			Yes [9]	RRM2, CP110, NuSAP1
Emerging tumor suppressor	FBXO4	FBXO4	Yes [12, 43]			Yes [10]	Cyclin D1, TRF1
Emerging oncogene	FBXO5 (Refer to Chap. 5 for more details)	FBXO5, Emi1	Yes [44]		Yes [45–47]		(Inhibitor of APC^{Cdh1} and APC^{Cdc20})
Undetermined	FBXO7	FBXO7	Yes [20]				CD43
Emerging tumor suppressor	FBXO32	FBXO32	Yes [31, 38]	Yes [33]		Yes (promoter methylation) [48]	eIF3-f, MyoD, MKP-1

Table 4.2 Summary of the mouse models for the FBXO class of F-Box proteins

F-box protein	Mouse model type						
	Knockout mouse model			Transgenic mouse model/phenotype			
	Whole-body knockout/phenotype	Tissue-specific knockout		Tissue-specific expression			
		Tissue	Phenotype	Tissue	Transgene	Phenotype	
FBXO1 Emerging tumor suppressor	*Fbxo1-/-*: embryonic lethal [8]	Various tissues	Normal [8]				
FBXO4 Emerging tumor suppressor	*Fbxo4+/-* or *Fbxo4-/-*: lymphomas, histiocytic sarcoma, mammary and hepatocellular carcinomas [43] *Fbxo4-/-*: viable and normal [12]						
FBXO7 Undetermined	*Fbxo7-/-*: increased pro-B cell and pro-erythroblast populations [20]						
FBXO32 Emerging tumor suppressor	*Fbxo32-/-*: resistance to atrophy and exaggerated cardiac hypertrophy [31, 38]			Heart	*a-MHC-Fbxo32*	Inhibition of cardiac hypertrophy [33]	

Fbxo1 exhibited cell cycle defects such as reduced cell cycle doubling time and delayed cell cycle reentry from quiescence, indicating that FBXO1 might play a pivotal role in cell cycle progression [8]. Currently, few studies have identified significant pathological evidence for FBXO1 function. To date, only one study identified low FBXO1 expression in hepatocellular carcinoma (HCC), which is associated with poor differentiation and unfavorable prognosis, indicating that down regulation of FBXO1 in HCC could act as a promising prognostic marker [9]. In line with these studies one might speculate that FBXO1 could function as an emerging tumor suppressor through the control of genome integrity by degradation of CP110, NuSAP1, and RRM2 during the S and G2 phases of the cell cycle.

4.2 Emerging Tumor Suppressor Role of FBXO4

FBXO4 (also known as FBX4) binds the chaperone αβ-crystallin as a cofactor for phosphorylation-dependent substrate recognition within the context of an active SCF E3 ubiquitin ligase. The SCF$^{FBXO4-\alpha\beta\text{-crystallin}}$ complex mediates the ubiquitination of Thr286-phosphorylated Cyclin D1 after its translocation to the cytoplasm (Table 4.3). Mutations in *FBXO4* that impair the ability of the encoded protein to

Table 4.3 Summary of identified ubiquitin substrates for the FBXO class of F-box proteins

F-box protein	Substrate	Signaling pathway/functions	References
FBXO1	RRM2	Genome integrity and DNA damage repair	[7]
Emerging tumor suppressor	CP110	Centrosome homeostasis and mitotic fidelity	[5]
	NuSAP1	Chromosome congression and segregation, and mitotic spindle organization	[6]
FBXO2	pre-integrin 1	ERAD pathway	[16]
Undetermined	SHPS-1	Cytoskeletal reorganization and cell–cell communication	[49]
	NR1	Neuronal development and information storage	[50]
	Connexin 26	Cochlear gap junction	[51, 52]
	UL9	Herpes simplex virus 1 replication-initiator protein	[53]
FBXO3	Fbxl2	An essential TRAF (tumor necrosis factor receptor-associated factor) inhibitor	[54]
Undetermined	HIPK2	A conserved serine/threonine nuclear kinase that interacts with homeodomain transcription factors	[55]
	P300	A transcriptional co-activator	[55]
FBXO4	Cyclin D1	G1/S-specific cyclin, cell cycle	[11, 56] [57] [58]
Emerging tumor suppressor			
	PIN2/TRF1	Telomere length and cell cycle	[59]

(continued)

Table 4.3 (continued)

F-box protein	Substrate	Signaling pathway/functions	References
FBXO6	Tcrα	T cell receptor, TCR signaling	[55]
Undetermined	Chk1	Serine/threonine-protein kinase required for checkpoint-mediated cell cycle arrest	[60, 61]
	DDOST	Oligosaccharyl transferase subunit	[62]
	RPN1	Oligosaccharyl transferase subunit	[62]
	RPN2	Oligosaccharyl transferase subunit	[62]
	STT3A	Oligosaccharyl transferase subunit	[62]
	STT3B	Oligosaccharyl transferase subunit	[62]
	ERLIN2	SPFH domain-containing protein 2	[62]
	ERO1L	Endoplasmic oxidoreductin-1-like protein	[62]
	GLA	Alpha-galactosidase A	[62]
	GLT25D1	Procollagen galactosyltransferase 1, Glycosyltransferase	[62]
	H6PD	GDH/6PGL endoplasmic bifunctional protein	[62]
	IKBIP	Inhibitor of NF kappa-B kinase-interacting protein	[62]
	LEPRE1	Prolyl 3-hydroxylase 1, Oxidoreductase	[62]
	NOMO2	Nodal modulator 2, Transmembrane helix	[62]
	ORP150	Hypoxia upregulated protein 1, Chaperone	[62]
	PIGS	GPI transamidase component PIG-S	[62]
	PLOD1	Procollagen-lysine,2-oxoglutarate 5-dioxygenase 1	[62]
	PLOD3	Procollagen-lysine,2-oxoglutarate 5-dioxygenase 3 membrane	[62]
	ATRN	Attractin, Inflammatory response	[62]
	IGF2R	Insulin-like growth factor 2 receptor	[62]
	PON2	Serum paraoxonase/arylesterase 2	[62]
	TFRC	Transferrin receptor protein 1 secreted	[62]
	CRTAP	Cartilage-associated protein	[62]
	GRN	Cytokine granulin	[62]
	LAMB2	Laminin subunit beta-2 Lysosome	[62]
	GLB1	Beta-galactosidase, Hydrolase	[62]
	GNS	N-Acetylglucosamine-6-sulfatase	[62]
	GUSB	Beta-glucuronidase	[62]
	NAGLU	Alpha-N-acetylglucosaminidase, Hydrolase	[62]
	PRCP	Lysosomal Pro-X carboxypeptidase nucleus	[62]
	LMNB1	Lamin-B1	[62]
FBXO7	CD43	A transmembrane cell surface protein, immune function	[20]
Undetermined			
FBXO9	Tel2	DNA damage response	[63]
Undetermined	Tti1	DNA damage response	[63]
FBXO8	Arf6	ADP ribosylation, endocytosis	[64]
Undetermined			

(continued)

Table 4.3 (continued)

F-box protein	Substrate	Signaling pathway/functions	References
FBXO10 Undetermined	Bcl2	Apoptosis	[65]
FBXO11 Undetermined	Cdt2	Cell cycle progression and genomic stability	[66–68]
	Bcl6	The product of a proto-oncogene implicated in the pathogenesis of human B-cell lymphomas	[69]
	p53	Tumor suppressor, transcription factor	[70]
FBXO15 Undetermined	P-glycoprotein/ ABCB1	ABC transporter	[71]
FBXO17 Undetermined	Arf1	Directing stress-induced gene expression in fission yeast	[72]
FBXO22 Undetermined	α-actinin	A cytoskeletal protein of the spectrin superfamily	[73]
	Filamin C	Muscle-specific filamin	[73]
	KDM4A	Histone demethylase	[74]
FBXO25 Undetermined	Nkx2-5	Heart formation and development	[75]
	Isl1	Insulin gene enhancer	[75]
	Hand1	Trophoblast giant cell differentiation and cardiac morphogenesis.	[75]
FBXO31 Undetermined	Par6c	Neuron polarity protein	[76]
	Cyclin D1	Cell cycle	[59]
FBXO32 Emerging tumor suppressor	MKP1	Phosphatase, JNK signaling, apoptosis	[36]
	eIF3-f	Translation, muscle protein synthesis	[34]
	MyoD	Txn factor, muscle cell differentiation	[77]
	BK-β(1)	Vascular large conductance Ca(2+)-activated K(+) (BK) channel accessory β(1) subunit	[78]
FBXO33 Undetermined	YB-1	Txn, translation, cell growth	[79]
FBXO40 Undetermined	IRS-1	An adaptor protein that is one of the major substrates of the insulin receptor kinase	[80]
FBXO44 Undetermined	BRCA1	DNA damage repair, E3 ligase	[81]
FBXO45 Undetermined	p73	Apoptosis and cell cycle	[82]
	Munc13-1	Neuron synaptic vesicle-priming factor	[83]

mediate Cyclin D1 ubiquitination have been detected in primary esophageal cancers [10]. In addition, impairment of the SCF^FBXO4-αβ-crystallin E3 ligase function by RNAi-mediated depletion of FBXO4 in some cell lines attenuated Cyclin D1 ubiquitination, resulting in Cyclin D1 accumulation and promotion of cell cycle progression [11]. However, recent genetic analysis did not observe such a change in Cyclin D1 stability or expression level in *Fbxo4*$^{-/-}$ mice suggesting that Cyclin D1 protein turnover regulation by the SCF^FBXO4-αβ-crystallin complex may be tissue or cellular context-dependent, or loss of FBXO4 might be compensated by another SCF complex in regard to Cyclin D stability [12]. FBXO4 also binds the telomere protein

TRF1 (telomeric repeat-binding factor 1) in a domain-dependent manner (Table 4.3). In the FBXO4-TRF1 co-crystal structure, FBXO4 adopts a GTPase-like fold that binds to TRF1 independently of covalent modifications [13, 14]. FBXO4 specifically targets non-telomeric TRF1 for degradation, and this regulation is imparted by the binding of TIN2 (TRF1-interacting nuclear protein 2) to TRF1 at telomeres, which physically blocks the FBXO4-binding site on TRF1.

Although *Fbxo4* null and heterozygote mice are viable without major developmental defects, both *Fbxo4*$^{+/-}$ and *Fbxo4*$^{-/-}$ mice develop tumors that include lymphomas, histiocytic sarcomas, and less frequently, mammary and hepatocellular carcinomas (Table 4.2). Mutations in *FBXO4* were identified in esophageal tumors and occur in the N-terminal regulatory region of FBXO4, a domain that is phosphorylated by 14-3-3 epsilon [15]. These mutations disrupt ligase dimerization and lead to reduction in SCF$^{FBXO4-a\beta-crystallin}$ E3 ligase activity, stabilization of its substrates and induction of malignant transformation. In addition, FBXO4 has an F-box associated domain (FBA) that is predicted to mediate binding to glycosylated substrates [16]. However, this is an underdeveloped area of F-box protein biology, and glycosylation-dependent substrates have yet to be identified for FBXO4. Taken together, these data suggest that FBXO4 might function as a tumor suppressor but additional studies are required to validate its physiological antitumor contributions.

4.3   The Possible Role of FBXO7 in Tumorigenesis Remains Largely Undetermined

FBXO7, an F-box protein without WD40 repeats or leucine-rich repeats (LRR), contains a C-terminal specific proline-rich region (PRR) that is important for substrate recognition [17, 18]. Recent studies have suggested that FBXO7 may function in a tissue-specific manner [17, 19–21]. Initially, FBXO7 was identified in immortalized fibroblasts as a putative proto-oncogene, in which it acts as a scaffold or chaperone to fold or stabilize the Cyclin D/Cdk6/p27 complex, thus activating Cdk6 and promoting cellular transformation, invasiveness, and tumor formation in nude mice [19]. Furthermore, FBXO7 protein levels were elevated in human lung and colon cancer biopsies compared with the normal tissues [19]. However, in hepatomas, FBXO7 was recognized as a tumor suppressor by degrading Cdk1/Cyclin B-phosphorylated HURP (hepatoma upregulated protein), a cell cycle-regulated oncogene associated with the mitotic spindle, and involved in growth control of human hepatocellular carcinoma [17] (Table 4.3). Consistent with a tumor suppressor function, in haematopoietic precursor cells, reduction of FBXO7 inversely correlated with CD43, and increased cell proliferation characterized by a shortened G1 phase [20]. In a study of primary haematopoietic stem and progenitor cells (HSPCs), the function of FBXO7 was largely linked with p53 status. In cells carrying intact p53, ectopic expression of FBXO7 suppressed colony formation capacity and altered the differentiation of HSPCs [21]. However, in the absence of p53, FBXO7 expression can promote T cell lymphomagenesis [21]. Yet much remains unknown about FBXO7, including the identification of FBXO7 substrates. Except for HURP, only cIAP1 and TRAF2 have been characterized as FBXO7 targets, which are involved in the negative regulation of NF-κB signaling through decreased RIP1 ubiquitination, leading to the resistance of TNFα signaling [18, 22] (Table 4.3). Mice carrying homozygous deletion of *Fbxo7* exhibited increased pro-B cells and pro-erythroblast populations, suggesting that FBXO7 could possibly promote maturation of precursor cells and have an antiproliferative function [20]. However, additional transgenic or compound knockout mouse models are required to decipher the physiological contribution of FBXO7 in tumorigenesis.

In addition to a role in cancer development, FBXO7 also was recognized as a bona fide Parkinson's disease (PD)-associated gene and designated as "*PARK15*" [23, 24]. Mutations in *FBXO7* have been identified in patients presenting with an early-onset form of PD [25, 26], which opened up new questions about the role of FBXO7 in preventing neurological diseases. Mutations in FBXO7 have been shown to decrease SCF^{FBXO7} E3 ligase activity by reducing FBXO7 affinity to Skp1 (such as the G877R mutant) [27] or blocking the interaction between FBXO7 and its substrates (such as the R498X mutant) [28]. Although these mutations established its E3 activity as important for neuronal health, PD-relevant substrates targeted by FBXO7 have yet to be identified.

FBXO7 also has been found in α-synuclein-positive Lewy bodies in idiopathic brain tissue and involved in Lewy body formation [29]. Further studies have found that FBXO7 regulates mitophagy through directly binding to and recruiting Parkin to depolarized mitochondria to initiate mitophagy. More importantly, restoration of human FBXO7 rescues phenotypes of Parkin loss in a *Drosophila* model of neurodegeneration [30], suggesting that the interaction of Parkin-FBXO7 is important for neuronal health. Moreover, FBXO7 also directly binds PINK1, and act as a scaffold to facilitate PINK1-mediated phosphorylation and activation of Parkin [30].

Taken together, FBXO7 has both SCF-dependent (with substrates HURP, cIAP, and TRAF2) and SCF-independent (enhance cyclin D/Cdk6/p21 complex formation, and PINK1/Parkin complex function) activities, which function in tissue-specific contexts, and is involved in cell cycle regulation, NF-κB pathway inhibition, and mitophagy initiation.

4.4   The Emerging Tumor Suppressor Role of FBXO32

FBXO32, also known as atrogin-1 or muscle atrophy F-box (MAFbx), was identified as a muscle-specific F-box protein that is specifically expressed in skeletal muscle cells and cardiomyocytes [31, 32]. FBXO32 negatively regulates myocyte cell size [33]. Importantly, upregulation of FBXO32 is attributed to skeletal muscle atrophy, while *Fbxo32* deficient skeletal muscle displays marked resistance to denervation atrophy [31]. SCF^{FBXO32} targets calcineurin, eIF3-f, MyoD, MAPK phosphatase-1 (MKP-1), and inhibitor of κB (IκB) for proteasome-dependent degradation to control muscle homeostasis [33–37] (Table 4.3).

Previous studies have identified an interplay between the Akt/FOXO signaling pathways and FBXO32. Specifically, FBXO32 enhances the activities of forkhead transcription factor FOXO1 and FOXO3a through K63-linked polyubiquitination, which blocks Akt-dependent physiological hypertrophy [38]. On the other hand, FOXO1 and FOXO3a function as activators of FBXO32 through transcriptional activation by binding a *cis*-element of *FBXO32* [39]. Therefore, in atrophic conditions in which Akt activity is low, FOXOs induce *FBXO32* mRNA expression level. That in turn increases SCF^{FBXO32}-dependent protein degradation in muscle, leading to subsequent muscle atrophy. In keeping with the notion, heart-specific *Fbxo32* overexpressing transgenic mice displayed enhanced FOXO transcriptional activities and inhibition of pathological cardiac hypertrophy [33]. Conversely, *Fbxo32* knock-out mice displayed resistance to atrophy and exaggerated cardiac hypertrophy in response to voluntary exercise [31, 38] (Table 4.2).

Recent findings have suggested that FBXO32 has a tumor suppressor function. Transcription of FBXO32 is suppressed in cancer through promoter methylation by EZH2, a catalytic subunit of the Polycomb Repressive Complex 2 (PRC2) [40, 41]. Inhibition of EZH2 activity induces *FBXO32* expression, thereby enhancing apoptosis of cancer cells partly through restoration of FBXO32 activity. Another study using an ovarian cancer model showed that ectopic expression of FBXO32 suppressed tumor growth in a xenograft mouse model by enhancing pro-apoptotic response of ovarian cancer cells, and sensitized chemo-resistant ovarian cancer cells to cisplatin [42]. Furthermore, an inverse correlation between the promoter methylation level of *FBXO32* and overall survival of ovarian cancer patients was observed [42]. These data suggest that restoration of FBXO32 is a potential therapeutic strategy against tumor growth, and a high methylation level of the *FBXO32* promoter could be a biomarker for cancer.

Taken together, FBXO32 plays critical roles in muscle homeostasis and cell growth inhibition, and dysregulation of FBXO32 may lead to several diseases such as pathological atrophy and cardiac hypertrophy, and cancer. Therefore, further studies are warranted to define the physiological roles of FBXO32 in these processes, and to understand the molecular mechanisms by which FBXO32 is involved in the development of these disorders.

References

1. Bai C, Richman R, Elledge SJ. Human cyclin F. EMBO J. 1994;13(24):6087–98.
2. Fung TK, et al. Cyclin F is degraded during G2-M by mechanisms fundamentally different from other cyclins. J Biol Chem. 2002;277(38):35140–9.
3. Kong M, et al. Cyclin F regulates the nuclear localization of cyclin B1 through a cyclin-cyclin interaction. EMBO J. 2000;19(6):1378–88.
4. Bai C, et al. SKP1 connects cell cycle regulators to the ubiquitin proteolysis machinery through a novel motif, the F-box. Cell. 1996;86(2):263–74.
5. D'Angiolella V, et al. SCF(Cyclin F) controls centrosome homeostasis and mitotic fidelity through CP110 degradation. Nature. 2010;466(7302):138–42.

6. Emanuele MJ, et al. Global identification of modular cullin-RING ligase substrates. Cell. 2011;147(2):459–74.
7. D'Angiolella V, et al. Cyclin F-mediated degradation of ribonucleotide reductase M2 controls genome integrity and DNA repair. Cell. 2012;149(5):1023–34.
8. Tetzlaff MT, et al. Cyclin F disruption compromises placental development and affects normal cell cycle execution. Mol Cell Biol. 2004;24(6):2487–98.
9. Fu J, et al. Low cyclin F expression in hepatocellular carcinoma associates with poor differentiation and unfavorable prognosis. Cancer Sci. 2013;104(4):508–15.
10. Barbash O, et al. Mutations in Fbx4 inhibit dimerization of the SCF(Fbx4) ligase and contribute to cyclin D1 overexpression in human cancer. Cancer Cell. 2008;14(1):68–78.
11. Lin DI, et al. Phosphorylation-dependent ubiquitination of cyclin D1 by the SCF(FBX4-alphaB crystallin) complex. Mol Cell. 2006;24(3):355–66.
12. Kanie T, et al. Genetic reevaluation of the role of F-box proteins in cyclin D1 degradation. Mol Cell Biol. 2012;32(3):590–605.
13. Li Y, Hao B. Structural basis of dimerization-dependent ubiquitination by the SCF(Fbx4) ubiquitin ligase. J Biol Chem. 2010;285(18):13896–906.
14. Zeng Z, et al. Structural basis of selective ubiquitination of TRF1 by SCFFbx4. Dev Cell. 2010;18(2):214–25.
15. Barbash O, Lee EK, Diehl JA. Phosphorylation-dependent regulation of SCF(Fbx4) dimerization and activity involves a novel component, 14-3-3varepsilon. Oncogene. 2011;30(17): 1995–2002.
16. Yoshida Y, et al. E3 ubiquitin ligase that recognizes sugar chains. Nature. 2002;418(6896): 438–42.
17. Hsu JM, et al. Fbx7 functions in the SCF complex regulating Cdk1-cyclin B-phosphorylated hepatoma up-regulated protein (HURP) proteolysis by a proline-rich region. J Biol Chem. 2004;279(31):32592–602.
18. Chang YF, et al. The F-box protein Fbxo7 interacts with human inhibitor of apoptosis protein cIAP1 and promotes cIAP1 ubiquitination. Biochem Biophys Res Commun. 2006;342(4): 1022–6.
19. Laman H, et al. Transforming activity of Fbxo7 is mediated specifically through regulation of cyclin D/cdk6. EMBO J. 2005;24(17):3104–16.
20. el Meziane K, et al. Knockdown of Fbxo7 reveals its regulatory role in proliferation and differentiation of haematopoietic precursor cells. J Cell Sci. 2011;124(Pt 13):2175–86.
21. Lomonosov M, et al. Expression of Fbxo7 in haematopoietic progenitor cells cooperates with p53 loss to promote lymphomagenesis. PLoS One. 2011;6(6):e21165.
22. Kuiken HJ, et al. Identification of F-box only protein 7 as a negative regulator of NF-kappaB signalling. J Cell Mol Med. 2012;16(9):2140–9.
23. Shojaee S, et al. Genome-wide linkage analysis of a Parkinsonian-pyramidal syndrome pedigree by 500 K SNP arrays. Am J Hum Genet. 2008;82(6):1375–84.
24. Klein C, Schneider SA, Lang AE. Hereditary parkinsonism: Parkinson disease look-alikes—an algorithm for clinicians to "PARK" genes and beyond. Mov Disord. 2009;24(14):2042–58.
25. Di Fonzo A, et al. FBXO7 mutations cause autosomal recessive, early-onset parkinsonian-pyramidal syndrome. Neurology. 2009;72(3):240–5.
26. Paisan-Ruiz C, et al. Early-onset L-dopa-responsive parkinsonism with pyramidal signs due to ATP13A2, PLA2G6, FBXO7 and spatacsin mutations. Mov Disord. 2010;25(12):1791–800.
27. Nelson DE, Laman H. A competitive binding mechanism between Skp1 and exportin 1 (CRM1) controls the localization of a subset of F-box proteins. J Biol Chem. 2011;286(22): 19804–15.
28. Zhao T, et al. Loss of nuclear activity of the FBXO7 protein in patients with parkinsonian-pyramidal syndrome (PARK15). PLoS One. 2011;6(2):e16983.
29. Zhao T, et al. FBXO7 immunoreactivity in alpha-synuclein-containing inclusions in Parkinson disease and multiple system atrophy. J Neuropathol Exp Neurol. 2013;72(6):482–8.
30. Burchell VS, et al. The Parkinson's disease-linked proteins Fbxo7 and Parkin interact to mediate mitophagy. Nat Neurosci. 2013;16(9):1257–65.

31. Bodine SC, et al. Identification of ubiquitin ligases required for skeletal muscle atrophy. Science. 2001;294(5547):1704–8.
32. Gomes MD, et al. Atrogin-1, a muscle-specific F-box protein highly expressed during muscle atrophy. Proc Natl Acad Sci U S A. 2001;98(25):14440–5.
33. Li HH, et al. Atrogin-1/muscle atrophy F-box inhibits calcineurin-dependent cardiac hypertrophy by participating in an SCF ubiquitin ligase complex. J Clin Invest. 2004;114(8):1058–71.
34. Csibi A, et al. MAFbx/Atrogin-1 controls the activity of the initiation factor eIF3-f in skeletal muscle atrophy by targeting multiple C-terminal lysines. J Biol Chem. 2009;284(7):4413–21.
35. Tintignac LA, et al. Degradation of MyoD mediated by the SCF (MAFbx) ubiquitin ligase. J Biol Chem. 2005;280(4):2847–56.
36. Xie P, et al. Atrogin-1/MAFbx enhances simulated ischemia/reperfusion-induced apoptosis in cardiomyocytes through degradation of MAPK phosphatase-1 and sustained JNK activation. J Biol Chem. 2009;284(9):5488–96.
37. Usui S, et al. Endogenous muscle atrophy F-box mediates pressure overload-induced cardiac hypertrophy through regulation of nuclear factor-kappaB. Circ Res. 2011;109(2):161–71.
38. Li HH, et al. Atrogin-1 inhibits Akt-dependent cardiac hypertrophy in mice via ubiquitin-dependent coactivation of Forkhead proteins. J Clin Invest. 2007;117(11):3211–23.
39. Sandri M, et al. Foxo transcription factors induce the atrophy-related ubiquitin ligase atrogin-1 and cause skeletal muscle atrophy. Cell. 2004;117(3):399–412.
40. Tan J, et al. Pharmacologic disruption of Polycomb-repressive complex 2-mediated gene repression selectively induces apoptosis in cancer cells. Genes Dev. 2007;21(9):1050–63.
41. Ciarapica R, et al. The Polycomb group (PcG) protein EZH2 supports the survival of PAX3-FOXO1 alveolar rhabdomyosarcoma by repressing FBXO32 (Atrogin1/MAFbx). Oncogene. Online publication 11 Nov 2013.
42. Chou JL, et al. Promoter hypermethylation of FBXO32, a novel TGF-beta/SMAD4 target gene and tumor suppressor, is associated with poor prognosis in human ovarian cancer. Lab Invest. 2010;90(3):414–25.
43. Vaites LP, et al. The Fbx4 tumor suppressor regulates cyclin D1 accumulation and prevents neoplastic transformation. Mol Cell Biol. 2011;31(22):4513–23.
44. Lee H, et al. Mouse emi1 has an essential function in mitotic progression during early embryogenesis. Mol Cell Biol. 2006;26(14):5373–81.
45. Chen JY, Wang MC, Hung WC. Bcr-Abl-induced tyrosine phosphorylation of Emi1 to stabilize Skp2 protein via inhibition of ubiquitination in chronic myeloid leukemia cells. J Cell Physiol. 2011;226(2):407–13.
46. Gutgemann I, et al. Emi1 protein accumulation implicates misregulation of the anaphase promoting complex/cyclosome pathway in ovarian clear cell carcinoma. Mod Pathol. 2008;21(4):445–54.
47. Lehman NL, et al. Oncogenic regulators and substrates of the anaphase promoting complex/cyclosome are frequently overexpressed in malignant tumors. Am J Pathol. 2007;170(5):1793–805.
48. Chou JL, et al. Promoter hypermethylation of FBXO32, a novel TGF-beta/SMAD4 target gene and tumor suppressor, is associated with poor prognosis in human ovarian cancer. Lab Invest. 2010;90(3):414–25.
49. Murai-Takebe R, et al. Ubiquitination-mediated regulation of biosynthesis of the adhesion receptor SHPS-1 in response to endoplasmic reticulum stress. J Biol Chem. 2004;279(12):11616–25.
50. Kato A, et al. Activity-dependent NMDA receptor degradation mediated by retrotranslocation and ubiquitination. Proc Natl Acad Sci U S A. 2005;102(15):5600–5.
51. Nelson RF, et al. Selective cochlear degeneration in mice lacking the F-box protein, Fbx2, a glycoprotein-specific ubiquitin ligase subunit. J Neurosci. 2007;27(19):5163–71.
52. Henzl MT, et al. The cochlear F-box protein OCP1 associates with OCP2 and connexin 26. Hear Res. 2004;191(1–2):101–9.

53. Eom CY, Lehman IR. Replication-initiator protein (UL9) of the herpes simplex virus 1 binds NFB42 and is degraded via the ubiquitin-proteasome pathway. Proc Natl Acad Sci U S A. 2003;100(17):9803–7.
54. Chen BB, et al. A combinatorial F box protein directed pathway controls TRAF adaptor stability to regulate inflammation. Nat Immunol. 2013;14(5):470–9.
55. Shima Y, et al. PML activates transcription by protecting HIPK2 and p300 from SCFFbx3-mediated degradation. Mol Cell Biol. 2008;28(23):7126–38.
56. Jia L, Sun Y. F-box proteins FBXO31 and FBX4 in regulation of cyclin D1 degradation upon DNA damage. Pigment Cell Melanoma Res. 2009;22(5):518–9.
57. Pontano LL, et al. Genotoxic stress-induced cyclin D1 phosphorylation and proteolysis are required for genomic stability. Mol Cell Biol. 2008;28(23):7245–58.
58. Lee TH, et al. The F-box protein FBX4 targets PIN2/TRF1 for ubiquitin-mediated degradation and regulates telomere maintenance. J Biol Chem. 2006;281(2):759–68.
59. Santra MK, Wajapeyee N, Green MR. F-box protein FBXO31 mediates cyclin D1 degradation to induce G1 arrest after DNA damage. Nature. 2009;459(7247):722–5.
60. Zhang YW, et al. The F box protein Fbx6 regulates Chk1 stability and cellular sensitivity to replication stress. Mol Cell. 2009;35(4):442–53.
61. Merry C, et al. Targeting the checkpoint kinase Chk1 in cancer therapy. Cell Cycle. 2010; 9(2):279–83.
62. Liu B, et al. Proteomic identification of common SCF ubiquitin ligase FBXO6-interacting glycoproteins in three kinds of cells. J Proteome Res. 2012;11(3):1773–81.
63. Fernandez-Saiz V, et al. SCFFbxo9 and CK2 direct the cellular response to growth factor withdrawal via Tel2/Tti1 degradation and promote survival in multiple myeloma. Nat Cell Biol. 2013;15(1):72–81.
64. Yano H, et al. Fbx8 makes Arf6 refractory to function via ubiquitination. Mol Biol Cell. 2008; 19(3):822–32.
65. Chiorazzi M, et al. Related F-box proteins control cell death in *Caenorhabditis elegans* and human lymphoma. Proc Natl Acad Sci U S A. 2013;110(10):3943–8.
66. Abbas T, et al. CRL1-FBXO11 promotes Cdt2 ubiquitylation and degradation and regulates Pr-Set7/Set8-mediated cellular migration. Mol Cell. 2013;49(6):1147–58.
67. Rossi M, et al. Regulation of the CRL4(Cdt2) ubiquitin ligase and cell-cycle exit by the SCF(Fbxo11) ubiquitin ligase. Mol Cell. 2013;49(6):1159–66.
68. Abbas T, Keaton M, Dutta A. Regulation of TGF-beta signaling, exit from the cell cycle, and cellular migration through cullin cross-regulation: SCF-FBXO11 turns off CRL4-Cdt2. Cell Cycle. 2013;12(14):2175–82.
69. Duan S, et al. FBXO11 targets BCL6 for degradation and is inactivated in diffuse large B-cell lymphomas. Nature. 2012;481(7379):90–3.
70. Abida WM, et al. FBXO11 promotes the Neddylation of p53 and inhibits its transcriptional activity. J Biol Chem. 2007;282(3):1797–804.
71. Katayama K, Noguchi K, Sugimoto Y. FBXO15 regulates P-glycoprotein/ABCB1 expression through the ubiquitin–proteasome pathway in cancer cells. Cancer Sci. 2013;104(6):694–702.
72. Lawrence CL, Jones N, Wilkinson CR. Stress-induced phosphorylation of *S. pombe* Atf1 abrogates its interaction with F box protein Fbh1. Curr Biol. 2009;19(22):1907–11.
73. Spaich S, et al. F-box and leucine-rich repeat protein 22 is a cardiac-enriched F-box protein that regulates sarcomeric protein turnover and is essential for maintenance of contractile function in vivo. Circ Res. 2012;111(12):1504–16.
74. Tan MK, Lim HJ, Harper JW. SCF(FBXO22) regulates histone H3 lysine 9 and 36 methylation levels by targeting histone demethylase KDM4A for ubiquitin-mediated proteasomal degradation. Mol Cell Biol. 2011;31(18):3687–99.
75. Jang JW, et al. A novel Fbxo25 acts as an E3 ligase for destructing cardiac specific transcription factors. Biochem Biophys Res Commun. 2011;410(2):183–8.
76. Vadhvani M, et al. The centrosomal E3 ubiquitin ligase FBXO31-SCF regulates neuronal morphogenesis and migration. PLoS One. 2013;8(2):e57530.

77. Jo C, Cho SJ, Jo SA. Mitogen-activated protein kinase kinase 1 (MEK1) stabilizes MyoD through direct phosphorylation at tyrosine 156 during myogenic differentiation. J Biol Chem. 2011;286(21):18903–13.
78. Lu T, et al. Reactive oxygen species signaling facilitates FOXO-3a/FBXO-dependent vascular BK channel beta1 subunit degradation in diabetic mice. Diabetes. 2012;61(7):1860–8.
79. Lutz M, et al. Proteasomal degradation of the multifunctional regulator YB-1 is mediated by an F-Box protein induced during programmed cell death. FEBS Lett. 2006;580(16):3921–30.
80. Shi J, et al. The SCF-Fbxo40 complex induces IRS1 ubiquitination in skeletal muscle, limiting IGF1 signaling. Dev Cell. 2011;21(5):835–47.
81. Lu Y, et al. The F-box protein FBXO44 mediates BRCA1 ubiquitination and degradation. J Biol Chem. 2012;287(49):41014–22.
82. Peschiaroli A, et al. The F-box protein FBXO45 promotes the proteasome-dependent degradation of p73. Oncogene. 2009;28(35):3157–66.
83. Tada H, et al. Fbxo45, a novel ubiquitin ligase, regulates synaptic activity. J Biol Chem. 2010;285(6):3840–9.

Chapter 5
The Role of APC E3 Ubiquitin Ligase Complex in Tumorigenesis

Jinfang Zhang, Lixin Wan, Brian J. North, Hiroyuki Inuzuka, and Wenyi Wei

Abstract E3 ubiquitin ligases, especially the SCF (Skp1-Cul1-F-box protein) and APC (Anaphase-promoting complex, also known as APC/C), have been extensively studied in the past decades. These two ubiquitin E3 ligases primarily function in the regulation of cell cycle progression through timely and coordinated degradation of key cell cycle regulators. Mounting evidence has revealed not only that the SCF plays a critical role in tumorigenesis but also that the APC is important for cancer development. Genetically modified mouse models have demonstrated that the APC co-activator Cdh1 primarily functions as a tumor suppressor, while another co-activator, Cdc20, exhibits an oncogenic role. Consistently, Cdh1 is frequently lost or inactivated in human cancers (Bassermann et al., Cell 134(2):256–267, 2008; Fujita et al., Clin Cancer Res 14(7):1966–1975, 2008; Fujita et al., Am J Pathol 173(1):217–228, 2008), while the overexpression of Cdc20 is observed in human malignancies (Mondal et al., Carcinogenesis 28(1):81–92, 2007; Jiang et al., Biochem Biophys Res Commun 415(2):325–329, 2011; Rajkumar et al., BMC Cancer 11:80, 2011; Chang et al., J Hematol Oncol 5:15, 2012; Kato et al., J Surg Oncol 106(4):423–430, 2012). Here, we discuss the identified substrates of APC, and summarize the reported phenotypes of genetically modified mouse models, which support the role of APC in the pathogenesis of human cancers and other relevant human diseases. Finally, we offer perspectives for developing APC pathway-specific inhibitors to treat various types of human cancers.

Keywords APC (Anaphase-promoting complex) • Cdh1 (Fzr1) • Cdc20 • Mitosis • SAC • Cell cycle • Development • D-box • KEN-box • Tumor suppressor • Oncoprotein • Mouse model • Physiological function

---

Jinfang Zhang and Lixin Wan have contributed equally to this chapter.

J. Zhang • L. Wan • B.J. North • H. Inuzuka • W. Wei (✉)
Department of Pathology, Beth Israel Deaconess Medical Center,
Harvard Medical School, Boston, MA 02215, USA
e-mail: wwei2@bidmc.harvard.edu

H. Inuzuka and W. Wei, *SCF and APC E3 Ubiquitin Ligases in Tumorigenesis*,
SpringerBriefs in Cancer Research, DOI 10.1007/978-3-319-05026-3_5,
© The Author(s) 2014

5.1 APC: Multi-subunit RING Finger Ubiquitin E3 Ligase

Two multisubunit RING finger ubiquitin E3 ligase families—the Skp1-CUL1-F-box protein (SCF) and the Anaphase-Promoting Complex (APC) have been shown to control cell cycle progression and has been implicated in a variety of human diseases, especially cancer [9, 10]. These two E3 ligases regulate cell cycle progression through timely and precise proteasome-dependent degradation of key cell cycle regulatory proteins such as cyclins and cyclin-dependent kinase (CDK) inhibitors [10]. The APC is composed of a constant core subunit, a Cullin-like subunit APC2, a RING finger protein APC11, and an interchangeable co-activator Cdc20 or Cdh1, which are mechanistically similar to the F-box proteins in the SCF E3 ligase complexes [11]. However, there are large differences between APC and SCF E3 ligase complexes in the complexity and diversity of their subunits and their physiological functions.

Currently, it is thought that at least 14 different proteins (namely, APC1/TSG24, APC2, APC3/Cdc27, APC4, APC5, APC6/Cdc6, APC7, APC8/Cdc23, APC10/Doc1, APC11, APC13/SWM1, APC15/Mnd2, APC16, and Cdc26) and one of the two co-activators (Cdc20 or Cdh1) form an large APC holoenzyme complex [12] (Fig. 5.1). Because of its large size and complex nature, the structure of APC holoenzyme is still undetermined. However, the general architecture of the complex is beginning to be understood though genetic, biochemical experiments and subunit structure analysis. A blueprint of the APC complex structure has begun to emerge showing that the APC consists of four subunits [12–14]—a scaffolding subunit, a catalytic and substrate recognition subunit, a tetratricopeptide repeat (TPR) arm,

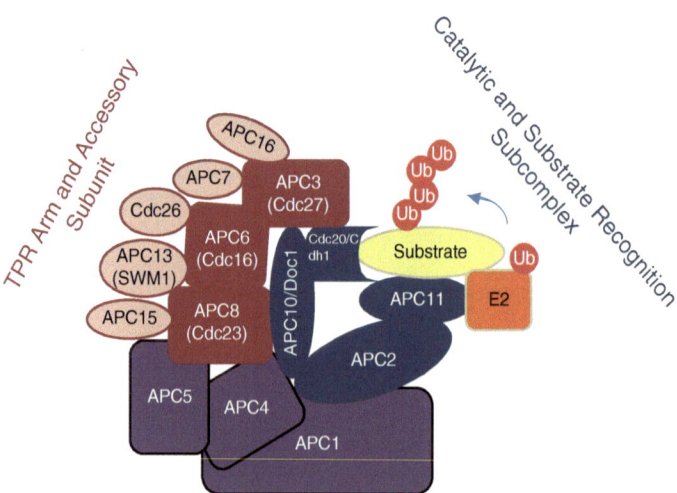

Fig. 5.1 Schematic illustration of the APC ubiquitin E3 ligase complex

and an accessory subunit (Fig. 5.1). The scaffolding subunit contains APC1, APC4, and APC5; the catalytic subunit is composed of APC2 (related to the Cullin protein family), APC11 (the RING finger protein) and APC10/Doc1; the TPR arm, which provides sites for binding the scaffolding unit and the co-activators (Cdc20 or Cdh1), consists of APC3/Cdc27, APC6/Cdc16, and APC8/Cdc23. In addition, the accessory subunits including Cdc26, APC13/Swm1, and the recently identified APC16, play an important role in stabilizing the TPR arm. Due to an important role in tumorigenesis [15–17], the APC is a potential therapeutic target for treating human cancer through designing small molecules for regulating APC activity through controlling its structural integrity, substrate recognition, or E3 ligase activity.

5.2 APC Core Subunit and Human Diseases

Although it has been reported that deletions and point mutations of several core subunits of the APC, such as APC3/Cdc27, APC6/Cdc16, and APC8/Cdc23, were found in colon cancer cell lines and human tumor samples [18], only two genetically modified mouse models with targeted inactivation of APC2 and APC10/Doc1 have been established (Table 5.1). The core component APC2 interacts with the RING finger domain protein APC11 to assemble a minimal APC catalytic core, which confirms that APC2 is critical for establishing the activity of the APC [19, 20]. To further address the physiological functions of APC2, an *Apc2* knockout mouse model was generated in 2004, which revealed that loss of *Apc2* led to early embryonic lethality at the E6.5 stage [21]. Notably, conditional depletion of *Apc2* in quiescent hepatocytes caused reentry into the cell cycle without any obvious external stimulus and subsequently arrested in metaphase [21]. Further analysis demonstrated that depletion of *Apc2* results in upregulated expression of several APC substrates such as Cyclin D1, Cyclin A2, Cyclin B1, Securin, and Plk in *Apc2*-deficient hepatocytes [21]. These results suggest that APC2 functions in cell cycle progression and early embryogenesis through degrading key cell cycle regulators such as mitotic cyclins and Securin. In addition, it was reported that the expression of *Apc2* is predominantly in neurons within the adult mouse brain [22]. Moreover, several APC core subunits and the co-activator Cdh1 were highly expressed in postmitotic, terminally differentiated neurons [23]. To uncover the physiological function of APC2 in the nervous system, an *Apc2*-conditional knockout mouse was utilized to specifically deplete *Apc2* in excitatory forebrain neurons [24]. Interestingly, the capacity to form spatial memories and extinction of fear memories were severely impaired in forebrain-specific *Apc2* knockout mice [24].

A mouse model carrying an *Apc10/Doc1*-null mutation was reported as early as the 1960s [25]. *Apc10/Doc1* was named *Oligosyndactylism* (*OS*) based on the phenotype of heterozygotic mice with fusion of the second and third digits on all four limbs [25, 26]. Further analysis showed that homozygotic *OS* mutants had an early and recessive embryonic lethality with a defect in the transition from metaphase to

Table 5.1 Summary of the knockout mouse models for the APC ubiquitin E3 ligase complex

APC subunits	Knockout mouse model		
		Tissue-specific knockout	
	Whole-body knockout/phenotype	Tissue	Phenotype
APC2	$Apc2^{-/-}$: Embryonic lethality before E6.5 [21]	Liver	Liver failure and premature death [21]
		Forebrain	Fail to extinct fear memories [24]
APC10/Doc1	$Apc10^{-/-}$: Early embryonic lethality [25, 27]		
	$Apc10^{+/-}$: Fusion of the second and third digits on all four limbs [28, 29]		
APC co-activators			
Cdc20 (oncogene)	$Cdc20^{-/-}$: Embryonic lethality at two-cell stage [34]		
	$Cdc20^{-/-}$ $Securin^{-/-}$: Embryonic lethality		
	Conditional $Cdc20^{-/-}$: Inhibiting the growth of tumors in vivo [32]		
	$Cdc20^{-/H}$: Females with either no or very few offspring [36]		
Cdh1 (tumor suppressor)	$Cdh1^{-/-}$: Embryonic lethality at E9.5–E10.5 stage [33]		
	$Cdh1^{+/-}$: Elevated spontaneous epithelial tumor incidence [33] and Learning/memory defects [39]		
APC inhibitors			
Mad2 (context-dependent roles in tumorigenesis)	$Mad2^{-/-}$: Embryonic lethality around E6.5–E7.5 [56]		
	$Mad2^{+/-}$: Prone to spontaneously develop lung tumors [57]		
	$Mad2^{+/-}$ $p53^{+/-}$: Higher tumor incidence relative to single mutant [58]		
Mad3/BubR1 (emerging role as a tumor suppressor)	$BubR1^{-/-}$: Embryonic lethality beyond E8.5 [75]		
	$BubR1^{+/-}$: Prone to develop tumors after AOM treatment [79]		
	$BubR1^{H/H}$: Prone to lung tumors after DMBA treatment, showed slow postnatal growth, infertility, and early onset of ageing phenotypes [76–78]		
	$BubR1^{+/-}$ adenomatous polyposis coli $(Apc)^{Min/+}$: Tenfold increase in colon tumors and a 50 % decrease in small intestine tumors [80]		
	$BubR1^{H/H}p16^{Ink4a-/-}$: Increased lung tumors relative and attenuated the age-related phenotypes [83]		
	$BubR1^{H/H}p19^{Arf-/-}$: Accelerated the ageing and no significant difference in survival curves relative to $p19^{Arf-/-}$ mice [83]		

(continued)

Table 5.1 (continued)

APC subunits	Knockout mouse model		
	Whole-body knockout/phenotype	Tissue-specific knockout	
		Tissue	Phenotype
Bub3 (emerging role as a tumor suppressor)	*Bub3⁻/⁻*: Embryonic lethality around E8.5 [85]		
	Bub3⁺/⁻: No obvious abnormalities in normal condition and have higher tumor incidence after DMBA treatment [85, 87]		
	Bub3⁺/⁻Trp53⁺/⁻: No substantial differences in survival or tumorigenesis [86]		
	Bub3⁺/⁻Rb1⁺/⁻: No significant differences in survival or tumorigenesis [86]		
	Bub3⁺/⁻Rae1⁺/⁻: Higher incidence of tumor formation after DMBA treatment [88]		
Emi1 (emerging role as an oncogene)	*Emi1⁻/⁻*: Embryonic lethality before implantation [95]		

+, wild type allele; –, null allele; H, hypomorphic allele; AOM, azoxymethane; DMBA, 7,12-dimethylbenz(α)anthracene

anaphase at the blastocyst stage, in part due to elevated stability of APC substrates including Cyclin B and Securin [27]. Subsequent analysis of the genomic locus surrounding the *OS* mutation identified that the anaphase-promoting complex core component *Apc10/Doc1* was disrupted, explaining the mitotic arrest phenotype of the radiation-induced *Os* mutant [28, 29]. Although expression of APC10/Doc1 was also detected in tissues that mostly contain differentiated cells, such as the adult brain [23, 30], tissue-specific knockout mouse models have yet to be established.

More importantly, to further understand the contribution of each APC core subunit to the E3 ligase activity of the APC, a systematic generation of genetic knockout mouse models targeting the remaining APC components will be a critical approach to decipher the physiological role of each APC core subunit in tumorigenesis. However, given the fact that germline knockout of either *Apc2* or *Apc10* causes embryonic lethality, tissue-specific conditional knockout strategies should be employed to generate these specific APC core component knockout mice, which will be of great interest in determining the causal role of these subunits in mammalian development and disease progression, including cancer.

5.3   APC Co-activators (Cdc20 and Cdh1) in Tumorigenesis

Compared to APC core components, the co-activators Cdc20 and Cdh1 have been relatively well-studied. Interestingly, although both Cdc20 and Cdh1 can activate the APC E3 ligase, they have opposing roles in regulating tumorigenesis. Mounting evidence has demonstrated that Cdc20 is frequently upregulated in human cancers including lung cancer [8], pancreatic ductal adenocarcinoma [7], cervical cancer [6], glioblastoma [31], oral cancer [4] and epithelial ovarian cancer [5]. Notably, depletion of Cdc20 inhibits tumor growth [32], which suggests that Cdc20 may function as an oncoprotein. Contrary to Cdc20, loss or low expression of Cdh1 is frequently observed in human tumors [1–3]. Correspondingly, knockdown of Cdh1 in many cancer cells retards their proliferation [33], implicating Cdh1 is a potential tumor suppressor. To understand physiological functions of these two co-activators in mammals, several genetic mouse models were generated by different research groups, which will be discussed in details below.

5.3.1 Cdc20 Functions as an Oncogene

In 2007, the Zhang group found that loss of *Cdc20* in mice through a gene-trapping method caused mouse embryonic lethality at the two-cell stage [34]. Further analysis demonstrated that the metaphase arrest depended on the function of Securin because *Cdc20* and *Securin* double mutant embryos rescued this phenotype [34] (Table 5.1). Subsequently, *Cdc20*-conditional knockout mice were generated and displayed an abundant metaphase arrest in proliferative tissues such as intestine and testis [32]. More importantly, this study has also demonstrated that depletion of *Cdc20* expression inhibited the growth of tumors in vivo with tumor cells arrested at metaphase-like stages and appearance of abundant apoptotic figures [32]. These results support a role for Cdc20 as an oncoprotein, consistent with its upregulation in a variety of human cancer tissues. Therefore, Cdc20 may facilitate human tumorigenesis and may represent a novel therapeutic target for treating human cancers.

Additionally, Cdc20 has also been implicated in controlling meiosis. Microinjecting specific anti-Cdc20 antibodies at pre-metaphase I caused a majority of oocytes to arrest at the metaphase I stage [35]. To further understand the physiological role of Cdc20 in meiosis and fertility, the van Deursen group generated a series of mutant mouse strains with graded reduction in *Cdc20* expression levels [36]. They found that these hypomorphic mice with low Cdc20 expression appeared healthy and had normal life span, folliculogenesis, and fertilization rates. However, the females produced either no, or very few, offspring. Furthermore, they discovered that a large portion of these embryos from hypomorphic female mice died during meiosis I in oocytes or after the first few embryogenic divisions [36]. Although these results demonstrated that a critical level of Cdc20 expression is essential for primary oocytes, whether Cdc20 insufficiency causes female infertility in humans remains to be determined.

5.3.2 Cdh1 Functions as a Tumor Suppressor

In contrast to the oncogenic role of Cdc20, emerging evidence has demonstrated that Cdh1 might play a tumor-suppressor role in human carcinogenesis.

Correspondingly, decreased expression of Cdh1 is observed in several tumor cell lines [37] and many tumor tissues, including prostate, ovary, breast, colon, liver, and brain [1–3]. Concomitantly, several APC^{Cdh1} targets including Cyclin B, Aurora A, Aurora B, Tpx2, Cdc6, and Cdc20 (Table 5.2), are frequently overexpressed in cancers with chromosomal instability [38]. Notably, genetically engineered mouse models further provide support that Cdh1 plays an important role in genomic stability and tumor suppression (Table 5.1). Specifically, knockout of *Cdh1* led to embryonic lethality at E9.5–E10.5 due to defective endoreduplication of trophoblast cells and placental malfunction [33]. Furthermore, *Cdh1* heterozygous mice exhibited decreased survival by 25 months of age and were susceptible to develop epithelial tumors, such as mammary adenocarcinomas and fibroadenomas [33]. These results suggest that Cdh1 functions as a haplo-insufficient tumor suppressor, which could explain the phenomenon that Cdh1 was downregulated in some human cancer cell lines and tissues. Subsequently, the Zhang group generated a *Cdh1* knockout mouse to study the importance of Cdh1 in vivo where they also observed early embryonic lethality around E9.5 [39]. Interestingly, they demonstrated that the *Cdh1* heterozygous mice were deficient in late-phase long-term potentiation (L-LTP) in the hippocampus [39]. However, developing a tissue-specific conditional knockout mouse model of *Cdh1* is needed to further identify the molecular functions of Cdh1 in human diseases such as cancer and neurodegeneration.

5.4   APC Substrates and Human Cancer

It has been reported that many APC substrates are aberrantly expressed in a variety of human cancers. Here, we summarize the reported substrates of APC and note their potential roles in tumorigenesis (Table 5.2). Increasing evidence has shown that APC^{Cdh1} has a tumor-suppressor role and downregulation of Cdh1 was observed in many tumor tissues [1–3]. Concomitantly, several APC^{Cdh1} substrates, such as Cyclin B, Aurora A, Aurora B, Plk1, Skp2, UbcH10, Tpx2, Cdc6, and Cdc20, are frequently upregulated in human cancers [16, 38]. Although Skp2 [40, 41] and UbcH10 [42, 43] have been characterized as specific APC^{Cdh1} substrates (Table 5.2), it remains elusive whether elevated expression of Plk1 and Aurora A are mainly due to defects in APC^{Cdh1} and/or APC^{Cdc20} activities. As shown in Table 5.2, it is plausible that unique substrates between Cdc20 and Cdh1 may be responsible for the different roles of Cdc20 and Cdh1 in tumorigenesis. APC^{Cdc20} typically degrades proteins containing a D-box [44], while APC^{Cdh1} could degrade a wider range of substrates containing either D-box [44], KEN-box [45], A-box [46, 47], O-box [48], CRY box [49], or GxEN box [43] motifs. Therefore, future studies should be directed to distinguish the overlapping or unique substrates of APC^{Cdh1} versus APC^{Cdc20} and their distinct roles in mediating the tumor-suppressor or oncogenic role of APC^{Cdh1} versus APC^{Cdc20}, respectively.

Table 5.2 Summary of the identified ubiquitin substrates for APC

Substrates	Functions/signaling pathways of substrates	Co-activators: Cdc20/Cdh1*	References
Aurora A	Mitotic serine/threonine kinases regulating cell cycle progression	Cdh1	[47]
Aurora B	Serine/threonine kinase controlling spindle assembly, chromosome alignment and segregation	Cdh1	[100, 101]
Anillin	An actin-binding protein required for cytokinesis	Cdh1	[102]
BRSK2	An AMP-activated protein kinase (AMPK)-related kinase	Cdh1	[103]
Bub1	Serine-threonine kinase that plays multiple roles in chromosome segregation and spindle checkpoint	Cdh1	[104]
Cdc5	Polo kinase Cdc5, a key factor in controlling Cdc14 localization	Cdh1	[105]
Cdc6	Governs the initiation of eukaryotic DNA replication	Cdh1	[106]
Cdc20	APC co-activator, recruiting substrates for APC-dependent degradation in early mitosis	Cdh1	[107, 108]
Cdc25A	Tyrosine protein phosphatase induces mitotic progression by dephosphorylating CDK1 and stimulating its kinase activity	Cdh1	[109]
Cdt1	Cooperates with CDC6 to initiate DNA replication	Cdh1	[110]
Cik1	Association with the kinesin Kar3 to control both the mitotic spindle and nuclear fusion during mating	Cdh1	[111]
CKAP2	A novel microtubule-associated protein that is frequently upregulated in various malignancies	Cdh1	[112, 113]
Cks1	The cofactor of Skp2 involving in G1–S transition	Cdh1	[40]
Claspin	Activates Chk1 and regulating DNA damage repair	Cdh1	[1, 114]
Ect2	GDP/GTP exchange factor regulates RhoA at mitosis	Cdh1	[115]
E2F3	Eukaryotic transcription factor and amplified in various human tumors	Cdh1	[116]
EYA1	Controls cell proliferation, survival and M–G1 transition	Cdh1	[117]
FAN1	FANCD2-associated nuclease 1, previously known as KIAA1018, is required for cellular resistance against DNA inter-strand cross-linking (ICL) agents	Cdh1	[118]
Fin1	A spindle-stabilizing protein in yeast	Cdh1	[119]

(continued)

Table 5.2 (continued)

Substrates	Functions/signaling pathways of substrates	Co-activators: Cdc20/Cdh1*	References
FoxM1	The forkhead box M1 (FoxM1) is a transcription factor that activates expression of the cell cycle genes required for both S and M phase progression	Cdh1	[120]
Geminin	Inhibits DNA replication by preventing the incorporation of MCM complex into pre-replication complex (pre-RC)	Cdh1	[121]
GluR1	Regulates homeostatic plasticity	Cdh1	[122]
GLS1	Metabolizes glutamine to glutamate	Cdh1	[123]
GLP	Regulates histone H3K9 methylation and senescence	Cdh1	[124]
G9a	Regulates histone H3K9 methylation and senescence	Cdh1	[124]
HEC1	Controls kinetochore microtubule dynamics and mitotic exit	Cdh1	[125]
Id2	Promotes axon growth	Cdh1	[126]
IQGAP	Promotes actomyosin-ring-independent cytokinesis at least in part by activation of Cyk3p in yeast	Cdh1	[127]
JNK	Controls cell survival, differentiation, and exit from mitosis	Cdh1	[128]
Kid	Also called KIF22, a Kinesin-like protein involved in spindle formation and the movements of chromosomes during mitosis and meiosis	Cdh1	[129]
Liprin-α	Regulates synaptic size	Cdh1	[130, 131]
MgcRacGAP	An important regulator of the Rho family GTPases-RhoA, Rac1, and Cdc42; indispensable in cytokinesis and cell cycle progression	Cdh1	[132]
MOAP-1	Modulator of apoptosis protein 1 (MOAP-1), an enhancer of Bax activation induced by DNA damage	Cdh1	[133]
Myf5	Muscle transcription factor	Cdh1	[134]
NEDL2	A HECT type ubiquitin ligase that enhances p73 transcriptional activity and degrades ATR kinase in lamin misexpressed cells	Cdh1	[135]
NIPA	An F-box-like protein that targets nuclear Cyclin B1 for degradation	Cdh1	[136]
Nrm1	Transcriptional activation of MBF in yeast	Cdh1	[137]
Pfkfb3	A key enzyme for regulating glycolysis	Cdh1	[138]
Plk1	A serine/threonine-protein kinase that activates MPF and assembles the mitotic spindle	Cdh1	[139]
Pr-set7	Protein methyltransferase playing an essential role in mammalian cell cycle progression	Cdh1	[140]
p190	Controls Rho activity and cell mobility	Cdh1	[141]

(continued)

Table 5.2 (continued)

Substrates	Functions/signaling pathways of substrates	Co-activators: Cdc20/Cdh1*	References
Rad17	Activates DNA damage checkpoint	Cdh1	[142]
RCS1	A mitotic regulator that controls the metaphase-to-anaphase transition	Cdh1	[143]
Sgo1	Shugoshin 1 (Sgo1) protects centromeric sister-chromatid cohesion in early mitosis, thus preventing premature sister-chromatid separation	Cdh1	[144]
Six1	An important mediator of normal development	Cdh1	[145]
SnoN	Inhibits TGF-β signaling and promote axon growth	Cdh1	[146]
Skp2	F-box protein that promotes the degradation of Cdk inhibitors p27^{Kip1} and p21^{Cip1}	Cdh1	[40, 41]
TACC3	Transforming acidic coiled-coil protein 3 (TACC3) is important for regulating mitotic spindle assembly and chromosome segregation	Cdh1	[147]
TK1	Thymidine kinase regulates dTTP production and genomic stability	Cdh1	[148, 149]
TMPK	Thymidylate kinase regulates dTTP production and genomic stability	Cdh1	[148]
Tpx2	Regulates spindle assembly	Cdh1	[150]
TRB3	Endoplasmic reticulum (ER) stress-inducible protein, is induced by CHOP and ATF4 to regulate their function and ER stress-induced cell death	Cdh1	[151]
UbcH10	E2 enzyme and essential factor of APC	Cdh1	[42]
UPS1	Ubiquitin carboxyl-terminal hydrolase 1 regulates DNA repair and genomic stability	Cdh1	[152]
Cenp-F	Required for kinetochore function and chromosome segregation in mitosis	Cdc20	[153]
Conductin	Inhibitor of the Wnt signaling pathway	Cdc20	[154]
Id1	Inhibits dendrite growth	Cdc20	[155]
Kif	Microtubule-depolymerizing kinesin that plays a role in chromosome congression	Cdc20	[156]
Mcl-1	Anti-apoptotic protein	Cdc20	[157]
Nek2A	Regulate centrosome separation and spindle formation	Cdc20	[158, 159]
NeuroD2	Neurogenic differentiation factor 2 inhibits presynaptic differentiation	Cdc20	[160]
PHF8	A demethylase PHF8 activates gene transcription primarily by demethylating histone H3 and H4	Cdc20	[161]
p21^{Cip1}	Inhibits cyclin-dependent kinase activity	Cdc20	[162]
REV1	Y-family polymerase specialized for replicating across DNA lesions at the stalled replication fork	Cdc20	[163]

(continued)

Table 5.2 (continued)

Substrates	Functions/signaling pathways of substrates	Co-activators: Cdc20/Cdh1*	References
Sp100	PML-NB scaffold protein, which localizes to nuclear particles during interphase and disperses from them during mitosis, participates in viral resistance, transcriptional regulation, and apoptosis	Cdc20	[164]
Bard1	A subunit of the Brca1-Bard1 tumor suppressor controlling spindle pole formation	Cdh1/Cdc20	[165]
Cyclin A	Controls S phase and G2–M transition	Cdh1/Cdc20	[166, 167]
Cyclin B	Activates Cdk1 and controls the G2–M transition	Cdh1/Cdc20	[168]
E2F1	Eukaryotic transcription factor to govern G1–S transition and apoptosis	Cdh1/Cdc20	[169]
Hmmr	Regulates the localization of Tpx2 at the spindle pole	Cdh1/Cdc20	[165]
HURP	Nucleates and cross-links microtubules in the vicinity of chromatin	Cdh1/Cdc20	[165]
Mps1	Dual specificity protein kinase with key roles in regulating the spindle assembly checkpoint and chromosome-microtubule attachments	Cdh1/Cdc20	[170]
Nlp	A key regulator in centrosome maturation that contributes to chromosome segregation and cytokinesis	Cdh1/Cdc20	[171]
NuSAP	Nucleates and cross-links microtubules in the vicinity of chromatin	Cdh1/Cdc20	[165]
RAP80	Recruits BRCA1 to DNA damage sites in the DNA damage-induced ubiquitin signaling pathway	Cdh1/Cdc20	[172]
Securin	Inhibits separase activity	Cdh1/Cdc20	[173, 174]
TRRAP	Histone acetyltransferase complex component	Cdh1/Cdc20	[175]

*Please note that the listed substrate as an APC/C^{Cdc20} and/or a APC/C^{Cdh1} target is based upon the original reports. Further validations are required to systematically determine whether any identified given substrate is a dual substrate for APC/C^{Cdc20} and APC/C^{Cdh1}, or a unique substrate for APC/C^{Cdc20} or APC/C^{Cdh1}

5.5   APC Endogenous Inhibitors in Tumorigenesis

To ensure proper cell cycle progression and genomic stability, the activity of the APC is tightly controlled by various mechanisms including endogenous inhibitor proteins (Tables 5.1 and 5.3). These natural inhibitors regulate the timely activity of APC to ensure cell cycle progression and chromosomal stability. Hence, these endogenous inhibitors have been shown to play an essential role in the development of human diseases, especially cancer [50, 51]. To identify the physiological functions of APC inhibitors in tumorigenesis, several genetic mouse models have been generated, which are discussed below.

Table 5.3 Summary of the transgenic mouse models for the APC ubiquitin E3 ligase complex

APC subunits	Transgenic mouse model					
	Whole-body expression/phenotype		Tissue-specific expression			
	Transgene	Phenotype	Tissue	Transgene	Phenotype	
APC inhibitors						
Mad2 (context-dependent roles in tumorigenesis)	*Tet-O Mad2*	A wide range of neoplasias [71]				
	Tet-O Mad2 and *Kras*	Accelerating lung tumorigenesis [72]				
Mad3/BubR1 (emerging role as a tumor suppressor)	Overexpression *BubR1*	Extended life span and delayed age-related deterioration and aneuploidy [84]				

+, wild type allele; −, null allele; Tet-O, tetracycline-inducible overexpression

5.5.1 Mitotic Arrest Deficient Protein 2 (Mad2) Displayed Context-Dependent Roles in Tumorigenesis

Mad2 inhibits APC^{Cdc20} through binding directly to Cdc20. However, robust inhibition requires subsequent binding of Mad2-Cdc20 to Mad3/BubR1 and Bub3, assembling what is known as the mitotic checkpoint complex (MCC) [52]. To ensure faithful chromosome segregation and genomic stability, the MCC blocks APC^{Cdc20}-mediated destruction of Cyclin B and Securin to delay the transition from metaphase to anaphase until all sister chromatids are properly aligned at the metaphase plate and attached to bipolar spindles [53–55]. To understand the physiological functions of Mad2 in vivo, a *Mad2* knockout mouse model was established in 2000 (Table 5.1). Loss of *Mad2* resulted in mouse embryonic lethality in utero exhibiting chromosome missegregation, marked decrease in the number of mitotic cells, and apoptotic death of a majority of embryos at E6.5–E7.5 [56]. Subsequently, the Benezra group demonstrated that deletion of one *Mad2* allele led to a defective mitotic checkpoint in both human cancer cells and murine embryonic fibroblasts with premature sister-chromatid separation and chromosome missegregation [57]. More importantly, they found that *Mad2* heterozygous mice were prone to developing papillary lung adenocarcinomas, an extremely rare tumor in wild-type mice [57]. Notably, *Mad2$^{+/-}$/p53$^{+/-}$* mice have higher cancer incidence and tumor burden than *p53$^{+/-}$* mice and the former are more prone to aneuploidy than the latter [58] (Table 5.1).

However, the role of Mad2 in tumorigenesis appears context-dependent as loss of heterozygosity in the region surrounding 4q27 containing *Mad2*, is observed in several tumor types [59–61]. Yet, overexpression of Mad2 is observed in many human cancers, including malignant lymphoma [62], lung cancer [63–65], liver cancer and hepatocellular carcinoma [66, 67], colorectal carcinoma [68], soft tissue sarcoma [69] and gastric cancer [70]. Recently, the Benezra lab has generated transgenic mice with conditional overexpression of Mad2 for exploring whether Mad2 upregulation contributes to tumorigenesis [71]. Transgenic mice overexpressing Mad2 resulted in a wide range of neoplasia, including lung adenomas, hepatocellular carcinomas, lymphomas, and fibrosarcomas [71]. Further analysis showed that *Mad2* overexpression resulted in stabilization of Securin and Cyclin B, delaying exit from mitosis, and increasing the incidence of aneuploid/polyploid cells [71]. Furthermore, Mad2-induced chromosome instability also led to lung tumor relapse in transgenic mice with tetracycline-inducible overexpression of KRAS and Mad2 after oncogenic KRAS withdrawal [72] (Table 5.3). These data support the notion that Mad2 overexpression causes a hyperactive mitotic checkpoint for overriding APC activity, which could explain the mitotic defects and chromosomal instability underlying tumorigenesis. However, mouse models with conditional knockout of *Mad2* and one gene of APC core components or activators would be useful to identify the precise molecular mechanism of tumorigenesis.

5.5.2 The Emerging Tumor Suppressor Role of Bub1-Related Protein (BubR1)

The mitotic checkpoint protein BubR1 (also named Mad3) inhibits the activity of APC^{Cdc20} through N-terminal binding of BubR1 to Mad2-Cdc20 and the APC core complex in a KEN box-dependent manner to inhibit the activity of APC in interphase, thereby permitting accumulation of Cyclin B prior to the onset of mitosis [73, 74]. Constitutive *BubR1* knockout in mice led to embryonic lethality beyond E8.5 in utero due to extensive apoptosis (Table 5.1). However, *BubR1$^{+/-}$* mice showed varied consequences and phenotypic characteristics compared to *Mad2$^{+/-}$* mice. *BubR1$^{+/-}$* mice had greater splenomegaly and extramedullary megakaryocytes, which was coupled with a decrease in erythroid, progenitors in bone marrow cells [75]. The observation of an increased number of megakaryocytes, a major cell type with a polyploid DNA content, indicated a slippage of mitotic arrest and abnormal mitotic exit in *BubR1$^{+/-}$* cells [75]. More importantly, the van Deursen group produced a series of mice with gradual reduction of BubR1 expression by using wild-type (*BubR1$^+$*), knockout (*BubR1$^-$*), and hypomorphic (*BubR1^H*) alleles [76] (Table 5.1). They demonstrated that *BubR1$^{-/-}$* mice died in early embryonic development and *BubR1$^{-/H}$* mice died after a few hours of birth. Although *BubR1$^{H/H}$* mice had normal phenotype at birth, they showed slow postnatal growth, infertility, and premature ageing phenotypes, with an increased prevalence of cellular senescence [76–78]. At the cellular level, lower levels of Cyclin B/Cdk1 kinase activity and pre-anaphase arrest in the presence of nocodazole were observed in mouse embryonic fibroblasts (MEFs) from *BubR1$^{H/H}$* animals [76].

Notably, Dai et al. found that *BubR1$^{+/-}$* mice were prone to rapid development of lung and colon adenocarcinomas after they have been exposed to carcinogen [79] (Table 5.1). Moreover, *BubR1$^{+/-}$* MEFs showed spontaneous micronuclei formation and lower levels of Securin, strongly suggesting that *BubR1* deficiency caused a compromised spindle checkpoint, which promoted the activity of APC [79]. Furthermore, the incidence of colonic tumors were increased tenfold by *BubR1* haplo-insufficiency in the adenomatous polyposis coli (*Apc^{Min/+}*) mice, accompanied by premature separation of sister chromatids in these compound mutant MEFs [80]. Interestingly, although expression levels of these two separate tumor suppressors, $p16^{Ink4a}$ and $p19^{Arf}$, were increased with age in different tissues [81, 82], depletion of these two tumor suppressors in *BubR1^{H/H}* mice exhibited opposing consequences and phenotypes [83]. Inactivation of *p16^{Ink4a}* attenuated the development of age-related pathological phenotypes; on the contrary, *p19^{Arf}* inactivation accelerated the process of ageing in *BubR1*-insufficient mice [83]. Moreover, *BubR1* insufficiency had synergistic effects on tumorigenesis in lung epithelial cells with loss of *p16^{Ink4a}*, but did not cooperate with *p19^{Arf}* loss to drive tumorigenesis [83]. Recently, Baker et al. reported that *BubR1* overexpressing transgenic mice protects against aneuploidy, extends healthy life span, and confer resistance to tumorigenesis, even in the presence of oncogenic *KRas^{G12D}* [84] (Table 5.3). These results from knockout and overexpression of *BubR1* demonstrated that this inhibitor plays an essential role in regulating chromosome stability and cancer development. However, why the synergistic effect of BubR1 insufficiency with $p16^{Ink4a}$ loss during tumorigenesis only occurs in lung epithelial cells, but not in other cell types remains largely unclear [83]. Thus, future studies using conditional knockout *BubR1* compounded with other related genes may be useful to explore molecular mechanisms of tissue-specific tumorigenesis.

5.5.3 The Emerging Tumor Suppressor Role of Budding Uninhibited by Benzimidazole Protein 3 (Bub3)

Bub3, a conserved constituent of the mitotic spindle assembly complex, plays an essential role in inhibiting the activity of APC, ensuring accurate segregation of chromosomes. To determine the physiological roles of Bub3 in vivo, *Bub3* knockout mice were generated in 2000 [85] (Table 5.1). *Bub3* null mice showed an embryonic lethal phenotype and died at E6.5–E7.5 in utero, resulting from the formation of micronuclei, chromatin bridging, lagging chromosomes, and grossly abnormal nuclear morphology [85]. However, *Bub3^{+/−}* mice had no obvious abnormalities in development or fertility compare to wild-type mice [85]. Moreover, there were no significant differences in the rates of survival or tumorigenesis between *Trp53^{+/−}* and *Bub3^{+/−}/Trp53^{+/−}* or between *Rb1^{+/−}* and *Bub3^{+/−}/Rb1^{+/−}* mice [86]. Importantly, another independent research group showed that although *Bub3^{+/−}* mice could not spontaneously develop tumors, these mutant mice were more prone to carcinogen-induced lung tumorigenesis than wild-type mice [87]. Furthermore, they found that combined haplo-insufficiency of *Bub3* and *Rae1* (*Bub3^{+/−}/Rae1^{+/−}*) accelerated chromosomal instability and the *Bub3^{+/−}/Rae1^{+/−}* mice had a higher incidence of tumor formation and average number of tumors per mouse after dimethylbenzanthrene treatment compared with single haplo-insufficient mice [88]. Together, these data suggest that the inhibitor Bub3 may function as a tumor-suppressor.

5.5.4 The Emerging Oncogenic Role of Early Mitotic Inhibitor 1 (Emi1 or FBXO5)

Emi1, also known as FBXO5, plays a crucial role in regulating the transition from G1 to S phase and mitotic progression through inhibiting the activity of APC. It has been reported that Emi1 is upregulated at the beginning of interphase and is ubiquitinated and degraded in early mitosis just before Cyclin A is destroyed by APC^{Cdc20} [89, 90]. Furthermore, the Pagano and Jackson groups demonstrated that $SCF^{\beta\text{-TRCP1}}$ could target Emi1 for degradation, which was required for timely activation of APC in early mitosis [91, 92]. The Pagano group generated $\beta\text{-}Trcp1$ knockout mice and found that Emi1 and mitotic Cyclins were accumulated in $\beta\text{-}Trcp1^{-/-}$ MEFs, which exhibited mitotic defects including multipolar metaphase spindles, misaligned chromosomes, and centrosome overduplication [91]. The Jackson group further revealed that degradation of Emi1 by $SCF^{\beta\text{-TRCP1}}$ was required for APC activation in early mitosis and failure of β-TRCP1-mediated Emi1 destruction caused stabilization of APC substrates and mitotic catastrophe, such as blockage of prometaphase, chromosome missegregation, and centrosome amplification [92]. Both groups showed that the DSGxxS motif in Emi1 and phosphorylation of the two serine residues in the motif were required for recruiting the $SCF^{\beta\text{-TRCP1}}$ ubiquitin E3 ligase to degrade Emi1 and permitting progression beyond prometaphase [91, 92]. Furthermore, polo-like kinase 1 (Plk1) could phosphorylate serine residues in the DSGxxS motif of Emi1 and markedly promote $SCF^{\beta\text{-TRCP1}}$ to bind and degrade Emi1 [93, 94]. To determine the physiological functions of Emi1 in vivo, Lee et al. generated $Emi1$-deficient mice using a gene-targeting technique [95]. They demonstrated that $Emi1^{-/-}$ embryos died at the preimplantation stage due to defects in mitotic progression, including multipolar spindle formation, chromosomal missegregation, and increased apoptosis [95]. Moreover, reduced level of Cyclin A was also observed in $Emi1$-deficient embryos [95].

Notably, Emi1 has been recently shown to have an oncogenic role in facilitating human tumorigenesis. Specifically, it was reported that overexpression of Emi1 is involved in regulating genomic instability of $p53$-deficient cells [96] and significant overexpression of Emi1 was observed in ovarian tumor tissues [97]. Moreover, the levels of Emi1 expression were associated with a high histologic grade and worse survival in ovarian cancer [97] and breast cancer [98]. Conversely, Emi1 depletion enhanced the sensitivity of doxorubicin or X-ray treatment in human cancer cells [99]. Altogether, these results demonstrated that Emi1 may function as an oncogenic role to promote human cancer development. However, whether Emi1 is involved in tumorigenesis under physiological conditions such as in genetically engineered mouse models remains largely undetermined in part due to the embryonic lethality. Therefore, further efforts in generating conditional *Emi1* knockout mice to pinpoint the physiological contribution of Emi1 in tumorigenesis are warranted.

References

1. Bassermann F, et al. The Cdc14B-Cdh1-Plk1 axis controls the G2 DNA-damage-response checkpoint. Cell. 2008;134(2):256–67.
2. Fujita T, et al. Dissection of the APCCdh1-Skp2 cascade in breast cancer. Clin Cancer Res. 2008;14(7):1966–75.
3. Fujita T, et al. Regulation of Skp2-p27 axis by the Cdh1/anaphase-promoting complex pathway in colorectal tumorigenesis. Am J Pathol. 2008;173(1):217–28.
4. Mondal G, et al. Overexpression of Cdc20 leads to impairment of the spindle assembly checkpoint and aneuploidization in oral cancer. Carcinogenesis. 2007;28(1):81–92.
5. Jiang J, Jedinak A, Sliva D. Ganodermanontriol (GDNT) exerts its effect on growth and invasiveness of breast cancer cells through the down-regulation of CDC20 and uPA. Biochem Biophys Res Commun. 2011;415(2):325–9.
6. Rajkumar T, et al. Identification and validation of genes involved in cervical tumourigenesis. BMC Cancer. 2011;11:80.
7. Chang DZ, et al. Increased CDC20 expression is associated with pancreatic ductal adenocarcinoma differentiation and progression. J Hematol Oncol. 2012;5:15.
8. Kato T, et al. Overexpression of CDC20 predicts poor prognosis in primary non-small cell lung cancer patients. J Surg Oncol. 2012;106(4):423–30.
9. Deshaies RJ, Joazeiro CA. RING domain E3 ubiquitin ligases. Annu Rev Biochem. 2009; 78:399–434.
10. Nakayama KI, Nakayama K. Ubiquitin ligases: cell-cycle control and cancer. Nat Rev Cancer. 2006;6(5):369–81.
11. Lipkowitz S, Weissman AM. RINGs of good and evil: RING finger ubiquitin ligases at the crossroads of tumour suppression and oncogenesis. Nat Rev Cancer. 2011;11(9):629–43.
12. Foe I, Toczyski D. Structural biology: a new look for the APC. Nature. 2011;470(7333): 182–3.
13. Schreiber A, et al. Structural basis for the subunit assembly of the anaphase-promoting complex. Nature. 2011;470(7333):227–32.
14. Vodermaier HC, et al. TPR subunits of the anaphase-promoting complex mediate binding to the activator protein CDH1. Curr Biol. 2003;13(17):1459–68.
15. Wasch R, Robbins JA, Cross FR. The emerging role of APC/CCdh1 in controlling differentiation, genomic stability and tumor suppression. Oncogene. 2010;29(1):1–10.
16. Penas C, Ramachandran V, Ayad NG. The APC/C ubiquitin ligase: from cell biology to tumorigenesis. Front Oncol. 2011;1:60.
17. Wang Z, et al. Cdc20: a potential novel therapeutic target for cancer treatment. Curr Pharm Des. 2013;19(18):3210–4.
18. Wang Q, et al. Alterations of anaphase-promoting complex genes in human colon cancer cells. Oncogene. 2003;22(10):1486–90.
19. Tang Z, et al. APC2 Cullin protein and APC11 RING protein comprise the minimal ubiquitin ligase module of the anaphase-promoting complex. Mol Biol Cell. 2001;12(12):3839–51.
20. Yu H, et al. Identification of a cullin homology region in a subunit of the anaphase-promoting complex. Science. 1998;279(5354):1219–22.
21. Wirth KG, et al. Loss of the anaphase-promoting complex in quiescent cells causes unscheduled hepatocyte proliferation. Genes Dev. 2004;18(1):88–98.

22. Yamanaka H, et al. Expression of Apc2 during mouse development. Brain Res Gene Expr Patterns. 2002;1(2):107–14.
23. Gieffers C, et al. Expression of the CDH1-associated form of the anaphase-promoting complex in postmitotic neurons. Proc Natl Acad Sci U S A. 1999;96(20):11317–22.
24. Kuczera T, et al. The anaphase promoting complex is required for memory function in mice. Learn Mem. 2011;18(1):49–57.
25. Van Valen P. Oligosyndactylism, an early embryonic lethal in the mouse. J Embryol Exp Morphol. 1966;15(2):119–24.
26. Stewart AD, Stewart J. Studies on syndrome of diabetes insipidus associated with oligosyndactyly in mice. Am J Physiol. 1969;217(4):1191–8.
27. Magnuson T, Epstein CJ. Oligosyndactyly: a lethal mutation in the mouse that results in mitotic arrest very early in development. Cell. 1984;38(3):823–33.
28. Pravtcheva DD, Wise TL. Disruption of Apc10/Doc1 in three alleles of oligosyndactylism. Genomics. 2001;72(1):78–87.
29. Pravtcheva DD, Wise TL. A transgene-induced mitotic arrest mutation in the mouse allelic with oligosyndactylism. Genetics. 1996;144(4):1747–56.
30. Almeida A, Bolanos JP, Moreno S. Cdh1/Hct1-APC is essential for the survival of postmitotic neurons. J Neurosci. 2005;25(36):8115–21.
31. Marucci G, et al. Gene expression profiling in glioblastoma and immunohistochemical evaluation of IGFBP-2 and CDC20. Virchows Arch. 2008;453(6):599–609.
32. Manchado E, et al. Targeting mitotic exit leads to tumor regression in vivo: modulation by Cdk1, Mastl, and the PP2A/B55alpha, delta phosphatase. Cancer Cell. 2010;18(6):641–54.
33. Garcia-Higuera I, et al. Genomic stability and tumour suppression by the APC/C cofactor Cdh1. Nat Cell Biol. 2008;10(7):802–11.
34. Li M, York JP, Zhang P. Loss of Cdc20 causes a securin-dependent metaphase arrest in two-cell mouse embryos. Mol Cell Biol. 2007;27(9):3481–8.
35. Yin S, et al. Cdc20 is required for the anaphase onset of the first meiosis but not the second meiosis in mouse oocytes. Cell Cycle. 2007;6(23):2990–2.
36. Jin F, et al. Cdc20 is critical for meiosis I and fertility of female mice. PLoS Genet. 2010;6(9):e1001147.
37. Engelbert D, et al. The ubiquitin ligase APC(Cdh1) is required to maintain genome integrity in primary human cells. Oncogene. 2008;27(7):907–17.
38. Carter SL, et al. A signature of chromosomal instability inferred from gene expression profiles predicts clinical outcome in multiple human cancers. Nat Genet. 2006;38(9):1043–8.
39. Li M, et al. The adaptor protein of the anaphase promoting complex Cdh1 is essential in maintaining replicative lifespan and in learning and memory. Nat Cell Biol. 2008;10(9):1083–9.
40. Bashir T, et al. Control of the SCF(Skp2-Cks1) ubiquitin ligase by the APC/C(Cdh1) ubiquitin ligase. Nature. 2004;428(6979):190–3.
41. Wei W, et al. Degradation of the SCF component Skp2 in cell-cycle phase G1 by the anaphase-promoting complex. Nature. 2004;428(6979):194–8.
42. Rape M, Kirschner MW. Autonomous regulation of the anaphase-promoting complex couples mitosis to S-phase entry. Nature. 2004;432(7017):588–95.
43. Castro A, et al. Xkid is degraded in a D-box, KEN-box, and A-box-independent pathway. Mol Cell Biol. 2003;23(12):4126–38.
44. Glotzer M, Murray AW, Kirschner MW. Cyclin is degraded by the ubiquitin pathway. Nature. 1991;349(6305):132–8.
45. Pfleger CM, Kirschner MW. The KEN box: an APC recognition signal distinct from the D box targeted by Cdh1. Genes Dev. 2000;14(6):655–65.
46. Castro A, et al. The D-Box-activating domain (DAD) is a new proteolysis signal that stimulates the silent D-Box sequence of Aurora-A. EMBO Rep. 2002;3(12):1209–14.
47. Littlepage LE, Ruderman JV. Identification of a new APC/C recognition domain, the A box, which is required for the Cdh1-dependent destruction of the kinase Aurora-A during mitotic exit. Genes Dev. 2002;16(17):2274–85.

48. Araki M, et al. Degradation of origin recognition complex large subunit by the anaphase-promoting complex in Drosophila. EMBO J. 2003;22(22):6115–26.
49. Reis A, et al. The CRY box: a second APCcdh1-dependent degron in mammalian cdc20. EMBO Rep. 2006;7(10):1040–5.
50. Schvartzman JM, Sotillo R, Benezra R. Mitotic chromosomal instability and cancer: mouse modelling of the human disease. Nat Rev Cancer. 2010;10(2):102–15.
51. Holland AJ, Cleveland DW. Boveri revisited: chromosomal instability, aneuploidy and tumorigenesis. Nat Rev Mol Cell Biol. 2009;10(7):478–87.
52. Sudakin V, Chan GK, Yen TJ. Checkpoint inhibition of the APC/C in HeLa cells is mediated by a complex of BUBR1, BUB3, CDC20, and MAD2. J Cell Biol. 2001;154(5):925–36.
53. Pinsky BA, Biggins S. The spindle checkpoint: tension versus attachment. Trends Cell Biol. 2005;15(9):486–93.
54. Yu H. Regulation of APC-Cdc20 by the spindle checkpoint. Curr Opin Cell Biol. 2002;14(6):706–14.
55. Musacchio A. Spindle assembly checkpoint: the third decade. Philos Trans R Soc Lond B Biol Sci. 2011;366(1584):3595–604.
56. Dobles M, et al. Chromosome missegregation and apoptosis in mice lacking the mitotic checkpoint protein Mad2. Cell. 2000;101(6):635–45.
57. Michel LS, et al. MAD2 haplo-insufficiency causes premature anaphase and chromosome instability in mammalian cells. Nature. 2001;409(6818):355–9.
58. Chi YH, et al. Spindle assembly checkpoint and p53 deficiencies cooperate for tumorigenesis in mice. Int J Cancer. 2009;124(6):1483–9.
59. Krishnan R, et al. Map location and gene structure of the Homo sapiens mitotic arrest deficient 2 (MAD2L1) gene at 4q27. Genomics. 1998;49(3):475–8.
60. Rashid A, et al. Genetic alterations in hepatocellular carcinomas: association between loss of chromosome 4q and p53 gene mutations. Br J Cancer. 1999;80(1–2):59–66.
61. Shivapurkar N, et al. Multiple regions of chromosome 4 demonstrating allelic losses in breast carcinomas. Cancer Res. 1999;59(15):3576–80.
62. Alizadeh AA, et al. Distinct types of diffuse large B-cell lymphoma identified by gene expression profiling. Nature. 2000;403(6769):503–11.
63. Garber ME, et al. Diversity of gene expression in adenocarcinoma of the lung. Proc Natl Acad Sci U S A. 2001;98(24):13784–9.
64. Heighway J, et al. Expression profiling of primary non-small cell lung cancer for target identification. Oncogene. 2002;21(50):7749–63.
65. Kato T, et al. Overexpression of MAD2 predicts clinical outcome in primary lung cancer patients. Lung Cancer. 2011;74(1):124–31.
66. Chen X, et al. Gene expression patterns in human liver cancers. Mol Biol Cell. 2002;13(6):1929–39.
67. Zhang SH, et al. Clinicopathologic significance of mitotic arrest defective protein 2 overexpression in hepatocellular carcinoma. Hum Pathol. 2008;39(12):1827–34.
68. Rimkus C, et al. Expression of the mitotic checkpoint gene MAD2L2 has prognostic significance in colon cancer. Int J Cancer. 2007;120(1):207–11.
69. Hisaoka M, Matsuyama A, Hashimoto H. Aberrant MAD2 expression in soft-tissue sarcoma. Pathol Int. 2008;58(6):329–33.
70. Wang L, et al. MAD2 as a key component of mitotic checkpoint: a probable prognostic factor for gastric cancer. Am J Clin Pathol. 2009;131(6):793–801.
71. Sotillo R, et al. Mad2 overexpression promotes aneuploidy and tumorigenesis in mice. Cancer Cell. 2007;11(1):9–23.
72. Sotillo R, et al. Mad2-induced chromosome instability leads to lung tumour relapse after oncogene withdrawal. Nature. 2010;464(7287):436–40.
73. Malureanu LA, et al. BubR1 N terminus acts as a soluble inhibitor of cyclin B degradation by APC/C(Cdc20) in interphase. Dev Cell. 2009;16(1):118–31.
74. Lara-Gonzalez P, et al. BubR1 blocks substrate recruitment to the APC/C in a KEN-box-dependent manner. J Cell Sci. 2011;124(Pt 24):4332–45.

75. Wang Q, et al. BUBR1 deficiency results in abnormal megakaryopoiesis. Blood. 2004; 103(4):1278–85.
76. Baker DJ, et al. BubR1 insufficiency causes early onset of aging-associated phenotypes and infertility in mice. Nat Genet. 2004;36(7):744–9.
77. Hartman TK, et al. Mutant mice with small amounts of BubR1 display accelerated age-related gliosis. Neurobiol Aging. 2007;28(6):921–7.
78. Matsumoto T, et al. Aging-associated vascular phenotype in mutant mice with low levels of BubR1. Stroke. 2007;38(3):1050–6.
79. Dai W, et al. Slippage of mitotic arrest and enhanced tumor development in mice with BubR1 haploinsufficiency. Cancer Res. 2004;64(2):440–5.
80. Rao CV, et al. Colonic tumorigenesis in BubR1+/−ApcMin/+ compound mutant mice is linked to premature separation of sister chromatids and enhanced genomic instability. Proc Natl Acad Sci U S A. 2005;102(12):4365–70.
81. Krishnamurthy J, et al. Ink4a/Arf expression is a biomarker of aging. J Clin Invest. 2004;114(9):1299–307.
82. Kim WY, Sharpless NE. The regulation of INK4/ARF in cancer and aging. Cell. 2006; 127(2):265–75.
83. Baker DJ, et al. Opposing roles for p16Ink4a and p19Arf in senescence and ageing caused by BubR1 insufficiency. Nat Cell Biol. 2008;10(7):825–36.
84. Baker DJ, et al. Increased expression of BubR1 protects against aneuploidy and cancer and extends healthy lifespan. Nat Cell Biol. 2013;15(1):96–102.
85. Kalitsis P, et al. Bub3 gene disruption in mice reveals essential mitotic spindle checkpoint function during early embryogenesis. Genes Dev. 2000;14(18):2277–82.
86. Kalitsis P, et al. Increased chromosome instability but not cancer predisposition in haploinsufficient Bub3 mice. Genes Chromosomes Cancer. 2005;44(1):29–36.
87. Babu JR, et al. Rae1 is an essential mitotic checkpoint regulator that cooperates with Bub3 to prevent chromosome missegregation. J Cell Biol. 2003;160(3):341–53.
88. Baker DJ, et al. Early aging-associated phenotypes in Bub3/Rae1 haploinsufficient mice. J Cell Biol. 2006;172(4):529–40.
89. Hsu JY, et al. E2F-dependent accumulation of hEmi1 regulates S phase entry by inhibiting APC(Cdh1). Nat Cell Biol. 2002;4(5):358–66.
90. Reimann JD, et al. Emi1 is a mitotic regulator that interacts with Cdc20 and inhibits the anaphase promoting complex. Cell. 2001;105(5):645–55.
91. Guardavaccaro D, et al. Control of meiotic and mitotic progression by the F box protein beta-Trcp1 in vivo. Dev Cell. 2003;4(6):799–812.
92. Margottin-Goguet F, et al. Prophase destruction of Emi1 by the SCF(betaTrCP/Slimb) ubiquitin ligase activates the anaphase promoting complex to allow progression beyond prometaphase. Dev Cell. 2003;4(6):813–26.
93. Hansen DV, et al. Plk1 regulates activation of the anaphase promoting complex by phosphorylating and triggering SCFbetaTrCP-dependent destruction of the APC inhibitor Emi1. Mol Biol Cell. 2004;15(12):5623–34.
94. Moshe Y, et al. Role of Polo-like kinase in the degradation of early mitotic inhibitor 1, a regulator of the anaphase promoting complex/cyclosome. Proc Natl Acad Sci U S A. 2004; 101(21):7937–42.
95. Lee H, et al. Mouse emi1 has an essential function in mitotic progression during early embryogenesis. Mol Cell Biol. 2006;26(14):5373–81.
96. Lehman NL, et al. Overexpression of the anaphase promoting complex/cyclosome inhibitor Emi1 leads to tetraploidy and genomic instability of p53-deficient cells. Cell Cycle. 2006;5(14):1569–73.
97. Min KW, et al. Clear cell carcinomas of the ovary: a multi-institutional study of 129 cases in Korea with prognostic significance of Emi1 and Galectin-3. Int J Gynecol Pathol. 2013;32(1): 3–14.
98. Liu X, et al. The expression and prognosis of Emi1 and Skp2 in breast carcinoma: associated with PI3K/Akt pathway and cell proliferation. Med Oncol. 2013;30(4):735.

99. Shimizu N, et al. Selective enhancing effect of early mitotic inhibitor 1 (Emi1) depletion on the sensitivity of doxorubicin or X-ray treatment in human cancer cells. J Biol Chem. 2013;288(24):17238–52.

100. Stewart S, Fang G. Destruction box-dependent degradation of aurora B is mediated by the anaphase-promoting complex/cyclosome and Cdh1. Cancer Res. 2005;65(19):8730–5.

101. Nguyen HG, et al. Mechanism of Aurora-B degradation and its dependency on intact KEN and A-boxes: identification of an aneuploidy-promoting property. Mol Cell Biol. 2005; 25(12):4977–92.

102. Zhao WM, Fang G. Anillin is a substrate of anaphase-promoting complex/cyclosome (APC/C) that controls spatial contractility of myosin during late cytokinesis. J Biol Chem. 2005;280(39):33516–24.

103. Li R, et al. APC/C(Cdh1) targets brain-specific kinase 2 (BRSK2) for degradation via the ubiquitin-proteasome pathway. PLoS One. 2012;7(9):e45932.

104. Qi W, Yu H. KEN-box-dependent degradation of the Bub1 spindle checkpoint kinase by the anaphase-promoting complex/cyclosome. J Biol Chem. 2007;282(6):3672–9.

105. Visintin C, et al. APC/C-Cdh1-mediated degradation of the Polo kinase Cdc5 promotes the return of Cdc14 into the nucleolus. Genes Dev. 2008;22(1):79–90.

106. Petersen BO, et al. Cell cycle- and cell growth-regulated proteolysis of mammalian CDC6 is dependent on APC-CDH1. Genes Dev. 2000;14(18):2330–43.

107. Huang JN, et al. Activity of the APC(Cdh1) form of the anaphase-promoting complex persists until S phase and prevents the premature expression of Cdc20p. J Cell Biol. 2001;154(1):85–94.

108. Hyun SY, et al. APC/C(Cdh1)-dependent degradation of Cdc20 requires a phosphorylation on CRY-box by Polo-like kinase-1 during somatic cell cycle. Biochem Biophys Res Commun. 2013;436(1):12–8.

109. Donzelli M, et al. Dual mode of degradation of Cdc25 A phosphatase. EMBO J. 2002; 21(18):4875–84.

110. Sugimoto N, et al. Identification of novel human Cdt1-binding proteins by a proteomics approach: proteolytic regulation by APC/CCdh1. Mol Biol Cell. 2008;19(3):1007–21.

111. Benanti JA, et al. Functionally distinct isoforms of Cik1 are differentially regulated by APC/C-mediated proteolysis. Mol Cell. 2009;33(5):581–90.

112. Seki A, Fang G. CKAP2 is a spindle-associated protein degraded by APC/C-Cdh1 during mitotic exit. J Biol Chem. 2007;282(20):15103–13.

113. Hong KU, et al. Functional importance of the anaphase-promoting complex-Cdh1-mediated degradation of TMAP/CKAP2 in regulation of spindle function and cytokinesis. Mol Cell Biol. 2007;27(10):3667–81.

114. Gao D, et al. Cdh1 regulates cell cycle through modulating the claspin/Chk1 and the Rb/E2F1 pathways. Mol Biol Cell. 2009;20(14):3305–16.

115. Liot C, et al. APC(cdh1) mediates degradation of the oncogenic Rho-GEF Ect2 after mitosis. PLoS One. 2011;6(8):e23676.

116. Ping Z, et al. APC/C (Cdh1) controls the proteasome-mediated degradation of E2F3 during cell cycle exit. Cell Cycle. 2012;11(10):1999–2005.

117. Sun J, et al. The phosphatase-transcription activator EYA1 is targeted by anaphase-promoting complex/Cdh1 for degradation at M-to-G1 transition. Mol Cell Biol. 2013;33(5):927–36.

118. Lai F, et al. Human KIAA1018/FAN1 nuclease is a new mitotic substrate of APC/C(Cdh1). Chin J Cancer. 2012;31(9):440–8.

119. Woodbury EL, Morgan DO. Cdk and APC activities limit the spindle-stabilizing function of Fin1 to anaphase. Nat Cell Biol. 2007;9(1):106–12.

120. Park HJ, et al. Anaphase-promoting complex/cyclosome-CDH1-mediated proteolysis of the forkhead box M1 transcription factor is critical for regulated entry into S phase. Mol Cell Biol. 2008;28(17):5162–71.

121. McGarry TJ, Kirschner MW. Geminin, an inhibitor of DNA replication, is degraded during mitosis. Cell. 1998;93(6):1043–53.

122. Fu AK, et al. APC(Cdh1) mediates EphA4-dependent downregulation of AMPA receptors in homeostatic plasticity. Nat Neurosci. 2011;14(2):181–9.
123. Colombo SL, et al. Molecular basis for the differential use of glucose and glutamine in cell proliferation as revealed by synchronized HeLa cells. Proc Natl Acad Sci U S A. 2011;108(52):21069–74.
124. Takahashi A, et al. DNA damage signaling triggers degradation of histone methyltransferases through APC/C(Cdh1) in senescent cells. Mol Cell. 2012;45(1):123–31.
125. Li L, et al. Anaphase-promoting complex/cyclosome controls HEC1 stability. Cell Prolif. 2011;44(1):1–9.
126. Lasorella A, et al. Degradation of Id2 by the anaphase-promoting complex couples cell cycle exit and axonal growth. Nature. 2006;442(7101):471–4.
127. Ko N, et al. Identification of yeast IQGAP (Iqg1p) as an anaphase-promoting-complex substrate and its role in actomyosin-ring-independent cytokinesis. Mol Biol Cell. 2007;18(12): 5139–53.
128. Gutierrez GJ, et al. Interplay between Cdh1 and JNK activity during the cell cycle. Nat Cell Biol. 2010;12(7):686–95.
129. Feine O, et al. Human Kid is degraded by the APC/C(Cdh1) but not by the APC/C(Cdc20). Cell Cycle. 2007;6(20):2516–23.
130. Teng FY, Tang BL. APC/C regulation of axonal growth and synaptic functions in postmitotic neurons: the Liprin-alpha connection. Cell Mol Life Sci. 2005;62(14):1571–8.
131. van Roessel P, et al. Independent regulation of synaptic size and activity by the anaphase-promoting complex. Cell. 2004;119(5):707–18.
132. Nishimura K, et al. APC(CDH1) targets MgcRacGAP for destruction in the late M phase. PLoS One. 2013;8(5):e63001.
133. Huang NJ, et al. The Trim39 ubiquitin ligase inhibits APC/CCdh1-mediated degradation of the Bax activator MOAP-1. J Cell Biol. 2012;197(3):361–7.
134. Doucet C, et al. Multiple phosphorylation events control mitotic degradation of the muscle transcription factor Myf5. BMC Biochem. 2005;6:27.
135. Lu L, et al. HECT type ubiquitin ligase NEDL2 is degraded by APC/C-Cdh1 and its tight regulation maintains the metaphase to anaphase transition. J Biol Chem. 2013;288(50): 35637–50.
136. Klitzing C, et al. APC/C(Cdh1)-mediated degradation of the F-box protein NIPA is regulated by its association with Skp1. PLoS One. 2011;6(12):e28998.
137. Ostapenko D, Solomon MJ. Anaphase promoting complex-dependent degradation of transcriptional repressors Nrm1 and Yhp1 in Saccharomyces cerevisiae. Mol Biol Cell. 2011;22(13):2175–84.
138. Herrero-Mendez A, et al. The bioenergetic and antioxidant status of neurons is controlled by continuous degradation of a key glycolytic enzyme by APC/C-Cdh1. Nat Cell Biol. 2009; 11(6):747–52.
139. Lindon C, Pines J. Ordered proteolysis in anaphase inactivates Plk1 to contribute to proper mitotic exit in human cells. J Cell Biol. 2004;164(2):233–41.
140. Wu S, et al. Dynamic regulation of the PR-Set7 histone methyltransferase is required for normal cell cycle progression. Genes Dev. 2010;24(22):2531–42.
141. Naoe H, et al. The anaphase-promoting complex/cyclosome activator Cdh1 modulates Rho GTPase by targeting p190 RhoGAP for degradation. Mol Cell Biol. 2010;30(16): 3994–4005.
142. Zhang L, et al. Proteolysis of Rad17 by Cdh1/APC regulates checkpoint termination and recovery from genotoxic stress. EMBO J. 2010;29(10):1726–37.
143. Zhao WM, et al. RCS1, a substrate of APC/C, controls the metaphase to anaphase transition. Proc Natl Acad Sci U S A. 2008;105(36):13415–20.
144. Karamysheva Z, et al. Multiple anaphase-promoting complex/cyclosome degrons mediate the degradation of human Sgo1. J Biol Chem. 2009;284(3):1772–80.
145. Christensen KL, et al. Cell cycle regulation of the human Six1 homeoprotein is mediated by APC(Cdh1). Oncogene. 2007;26(23):3406–14.

146. Stegmuller J, et al. Cell-intrinsic regulation of axonal morphogenesis by the Cdh1-APC target SnoN. Neuron. 2006;50(3):389–400.
147. Jeng JC, et al. Cdh1 controls the stability of TACC3. Cell Cycle. 2009;8(21):3529–36.
148. Ke PY, et al. Control of dTTP pool size by anaphase promoting complex/cyclosome is essential for the maintenance of genetic stability. Genes Dev. 2005;19(16):1920–33.
149. Ke PY, et al. Hiding human thymidine kinase 1 from APC/C-mediated destruction by thymidine binding. FASEB J. 2007;21(4):1276–84.
150. Stewart S, Fang G. Anaphase-promoting complex/cyclosome controls the stability of TPX2 during mitotic exit. Mol Cell Biol. 2005;25(23):10516–27.
151. Ohoka N, et al. Anaphase-promoting complex/cyclosome-cdh1 mediates the ubiquitination and degradation of TRB3. Biochem Biophys Res Commun. 2010;392(3):289–94.
152. Cotto-Rios XM, et al. APC/CCdh1-dependent proteolysis of USP1 regulates the response to UV-mediated DNA damage. J Cell Biol. 2011;194(2):177–86.
153. Gurden MD, et al. Cdc20 is required for the post-anaphase, KEN-dependent degradation of centromere protein F. J Cell Sci. 2010;123(Pt 3):321–30.
154. Hadjihannas MV, et al. Cell cycle control of Wnt/beta-catenin signalling by conductin/axin2 through CDC20. EMBO Rep. 2012;13(4):347–54.
155. Kim AH, et al. A centrosomal Cdc20-APC pathway controls dendrite morphogenesis in post-mitotic neurons. Cell. 2009;136(2):322–36.
156. Sedgwick GG, et al. Mechanisms controlling the temporal degradation of Nek2A and Kif18A by the APC/C-Cdc20 complex. EMBO J. 2013;32(2):303–14.
157. Harley ME, et al. Phosphorylation of Mcl-1 by CDK1-cyclin B1 initiates its Cdc20-dependent destruction during mitotic arrest. EMBO J. 2010;29(14):2407–20.
158. Hayes MJ, et al. Early mitotic degradation of Nek2A depends on Cdc20-independent interaction with the APC/C. Nat Cell Biol. 2006;8(6):607–14.
159. Hames RS, et al. APC/C-mediated destruction of the centrosomal kinase Nek2A occurs in early mitosis and depends upon a cyclin A-type D-box. EMBO J. 2001;20(24):7117–27.
160. Yang Y, et al. A Cdc20-APC ubiquitin signaling pathway regulates presynaptic differentiation. Science. 2009;326(5952):575–8.
161. Lim HJ, et al. The G2/M regulator histone demethylase PHF8 is targeted for degradation by the anaphase-promoting complex containing CDC20. Mol Cell Biol. 2013;33(21):4166–80.
162. Amador V, et al. APC/C(Cdc20) controls the ubiquitin-mediated degradation of p21 in prometaphase. Mol Cell. 2007;27(3):462–73.
163. Chun AC, Kok KH, Jin DY. REV7 is required for anaphase-promoting complex-dependent ubiquitination and degradation of translesion DNA polymerase REV1. Cell Cycle. 2013;12(2):365–78.
164. Wang R, et al. Cdc20 mediates D-box-dependent degradation of Sp100. Biochem Biophys Res Commun. 2011;415(4):702–6.
165. Song L, Rape M. Regulated degradation of spindle assembly factors by the anaphase-promoting complex. Mol Cell. 2010;38(3):369–82.
166. den Elzen N, Pines J. Cyclin A is destroyed in prometaphase and can delay chromosome alignment and anaphase. J Cell Biol. 2001;153(1):121–36.
167. Geley S, et al. Anaphase-promoting complex/cyclosome-dependent proteolysis of human cyclin A starts at the beginning of mitosis and is not subject to the spindle assembly checkpoint. J Cell Biol. 2001;153(1):137–48.
168. Clute P, Pines J. Temporal and spatial control of cyclin B1 destruction in metaphase. Nat Cell Biol. 1999;1(2):82–7.
169. Budhavarapu VN, et al. Regulation of E2F1 by APC/C Cdh1 via K11 linkage-specific ubiquitin chain formation. Cell Cycle. 2012;11(10):2030–8.
170. Cui Y, et al. Degradation of the human mitotic checkpoint kinase Mps1 is cell cycle-regulated by APC-cCdc20 and APC-cCdh1 ubiquitin ligases. J Biol Chem. 2010;285(43):32988–98.
171. Wang Y, Zhan Q. Cell cycle-dependent expression of centrosomal ninein-like protein in human cells is regulated by the anaphase-promoting complex. J Biol Chem. 2007;282(24):17712–9.

172. Cho HJ, et al. Degradation of human RAP80 is cell cycle regulated by Cdc20 and Cdh1 ubiquitin ligases. Mol Cancer Res. 2012;10(5):615–25.
173. Michaelis C, Ciosk R, Nasmyth K. Cohesins: chromosomal proteins that prevent premature separation of sister chromatids. Cell. 1997;91(1):35–45.
174. Nasmyth K. Disseminating the genome: joining, resolving, and separating sister chromatids during mitosis and meiosis. Annu Rev Genet. 2001;35:673–745.
175. Ichim G, et al. The histone acetyltransferase component TRRAP is targeted for destruction during the cell cycle. Oncogene. 2014;33(2):181–92.

Chapter 6
Conclusions and Research Perspectives

Pengda Liu, Brian J. North, Hiroyuki Inuzuka, and Wenyi Wei

Abstract The SCF and APC ubiquitin E3 ligases are well-characterized members of the RING finger ubiquitin E3 ligase family. During the past two decades, a significant amount of experimental evidence has been generated towards understanding the biochemical features and the physiological functions of these ligases, especially their roles in tumorigenesis. In this book, we have discussed the growing list of the identified ubiquitin substrates for SCF and APC members. More importantly, we have summarized the studies that used genetically engineered mouse models to examine the roles of SCF or APC in tumorigenesis. Moreover, we have discussed the physiological functions of the SCF and APC members under normal conditions as well as in human diseases such as cancer. However, as there are numerous questions remaining to be addressed in this field, we urge the development of novel large-scale screening strategies to efficiently identify novel substrates for each E3 ligase, and to establish additional genetically engineered mouse models, both of which would greatly facilitate our understanding of the physiological as well as pathological roles of SCF and APC in tumorigenesis (Fig. 6.1).

Keywords APC • SCF • Mouse model • Perspective • Future directions • APC inhibitor • Skp2 inhibitor • Suppressing F-box protein or signaling pathway • Drug target • Cancer therapy

To date, a majority of research efforts have been exclusively focused on elucidating the roles of three well-characterized F-box proteins, Fbw7, Skp2, and β-TRCP, in tumorigenesis. However, recent studies have begun to unravel the functions of other

P. Liu • B.J. North • H. Inuzuka (✉) • W. Wei (✉)
Department of Pathology, Beth Israel Deaconess Medical Center,
Harvard Medical School, Boston, MA 02215, USA
e-mail: hinuzuka@bidmc.harvard.edu; wwei2@bidmc.harvard.edu

H. Inuzuka and W. Wei, *SCF and APC E3 Ubiquitin Ligases in Tumorigenesis*,
SpringerBriefs in Cancer Research, DOI 10.1007/978-3-319-05026-3_6,
© The Author(s) 2014

less understood F-box proteins through genetic, biochemical, and cell biological approaches. To this end, in vivo studies with genetically engineered mouse models have provided important insights, and will also further promote our understanding of the physiological functions of the remaining 66 putative SCF E3 ligases. Given that many F-box proteins might play critical roles in development or tissue homeostasis such that genetic deletion of a given F-box protein often leads to embryonic lethality, it is highly suggested that conditional knockout strategies should be utilized to systematically delete each F-box protein to understand whether loss of any given F-box protein facilitates tumorigenesis. Conversely, utilizing Tetracycline (Tet)-inducible F-box protein transgenic mouse strains will help to decipher possible oncogenic roles for any of the 69 putative F-box proteins in vivo.

Moreover, with APC being a large complex consisting of many subunits, it is somewhat surprising that only knockout mouse models targeting either the *APC2* or the *APC10/Doc1* APC core subunit in addition to targeted deletion of Cdh1 or Cdc20 adapter proteins, have been reported. Thus, it remains elusive why the APC complex requires so many subunits to form a functional E3 ubiquitin ligase. Therefore, further mouse models targeting the genetic ablation of each of the remaining 12 APC core subunits would be critical for revealing their contribution to APC E3 ligase activity as well as their physiological roles in governing tumorigenesis. Undoubtedly, these genetic mouse modeling results should be coupled with additional biochemical characterization of each subunit under physiological and pathological conditions to greatly facilitate our understanding of the physiological functions of each APC core subunit. Taken together, these future studies will ultimately provide novel insights into the physiological significance of the SCF and APC complexes and their potential roles in tumorigenesis.

Furthermore, recent progress in molecular oncology, especially the rapid advances in genomic and proteomic approaches, has made it readily possible to systematically link various gene alterations or epigenetic aberrations, and deregulation of intracellular signaling pathways with multiplex properties of many human cancers. For instance, Skp2 and Cdc20 overexpression or Fbw7 deletion/missense mutation are critical hallmarks in various types of human cancers. Additional studies studying SCF and APC components that have gene amplification, deletion, mutations, promoter methylation as well as aberrant protein modifications which provoke oncogenic signaling pathways in cancers will further provide many biological and clinical insights into the roles of SCF and APC complexes in tumorigenesis.

One of the primary reasons for elucidating the functions of SCF or APC in tumorigenesis is to design novel therapeutic approaches targeting these two critical E3 ubiquitin ligase complexes for cancer treatment (Fig. 6.1). Recent studies have demonstrated that conjugation of the ubiquitin-like molecule NEDD8 to the Cullin subunit of the CRL is critical for the E3 ligase complex to adopt a catalytically active conformation [1–3]. As a result, a recently developed NEDD8 E1-specific inhibitor MLN4924 [4] that directly inhibits the neddylation of Cullins, has been shown to be a novel class of anticancer agent. Furthermore, a recent study has demonstrated that the newly developed small molecule CC0651, which specifically

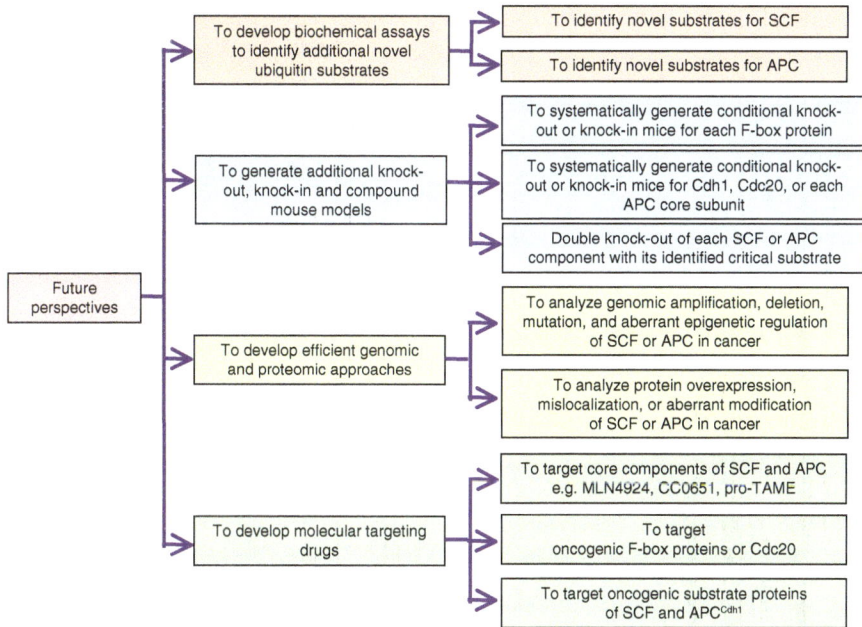

Fig. 6.1 Future perspectives to further understand the contributions of SCF and APC E3 ubiquitin ligase complexes in tumorigenesis

inhibits the E2 conjugating enzyme UBE2R1, could be another promising anticancer drug as UBE2R1 selectively functions with the CRL type E3 ligases [5]. However, it is expected that these compounds also inhibit the degradation of many other substrates targeted by CRLs or UBE2R1, and therefore, more specific strategies that target individual F-box proteins or APC activators are more desirable. For example, Skp2 has been well characterized as an oncoprotein and its overexpression is a hallmark of various human cancers including breast and prostate cancers [6, 7]. Thus, it is conceivable that direct targeting oncogenic E3 ligase complexes such as SCF^{Skp2} [8, 9], or targeting oncoproteins whose upstream tumor suppressive E3 function is compromised in cancer, could be a powerful strategy.

In this regard, Fbw7 is considered as a tumor suppressor and loss of Fbw7 has been implicated in cellular resistance to Taxol [10]. Therefore, either development of inhibitors for upstream regulatory proteins to activate Fbw7, or inhibiting its downstream oncoprotein targets would be viable therapeutic approaches. More importantly, there are F-box proteins such as β-TRCP, which displays context-dependent roles in tumorigenesis that requires more thorough examination of the mechanisms dictating their roles in tumorigenesis in a cancer type-specific or tissue context-dependent manner, to effectively design targeted therapeutic strategies based on their E3 ligase activity. To this end, it is reasonable to design personalized medicine based on the unique SCF or APC signaling signatures as efficient anticancer treatment by aiming at the specific signaling component along the SCF or APC pathways. In addition, while the APC inhibitor pro-TAME has been shown to induce spindle checkpoint-dependent mitotic arrest [11], the emerging opposing roles for Cdh1 and Cdc20 in tumorigenesis strengthens the point that development of a specific inhibitor targeting APC^{Cdc20} might be a more efficient intervention. To this end, additional biochemical analysis to further dissect APC^{Cdh1} substrates from APC^{Cdc20} substrates would also benefit our understanding of the opposing roles for APC^{Cdh1} and APC^{Cdc20} in turmorigenesis.

Taken together, the rapid increase in the basic and translational research in the field of cancer biology, particularly targeting the proteasome-dependent protein degradation pathways, will provide many useful clues to develop novel strategies for diagnosing, treating, and preventing cancer.

References

1. Chiba T, Tanaka K. Cullin-based ubiquitin ligase and its control by NEDD8-conjugating system. Curr Protein Pept Sci. 2004;5(3):177–84.
2. Read MA, et al. Nedd8 modification of cul-1 activates SCF(beta(TrCP))-dependent ubiquitination of IkappaBalpha. Mol Cell Biol. 2000;20(7):2326–33.
3. Podust VN, et al. A Nedd8 conjugation pathway is essential for proteolytic targeting of p27Kip1 by ubiquitination. Proc Natl Acad Sci U S A. 2000;97(9):4579–84.
4. Soucy TA, et al. An inhibitor of NEDD8-activating enzyme as a new approach to treat cancer. Nature. 2009;458(7239):732–6.
5. Ceccarelli DF, et al. An allosteric inhibitor of the human Cdc34 ubiquitin-conjugating enzyme. Cell. 2011;145(7):1075–87.
6. Wang Z, et al. Skp2 is a promising therapeutic target in breast cancer. Front Oncol. 2012;1(57): pii:18702.
7. Wang Z, et al. Skp2: a novel potential therapeutic target for prostate cancer. Biochim Biophys Acta. 2012;1825(1):11–7.
8. Chan CH, et al. Pharmacological inactivation of Skp2 SCF ubiquitin ligase restricts cancer stem cell traits and cancer progression. Cell. 2013;154(3):556–68.
9. Wu L, et al. Specific small molecule inhibitors of Skp2-mediated p27 degradation. Chem Biol. 2012;19(12):1515–24.
10. Wertz IE, et al. Sensitivity to antitubulin chemotherapeutics is regulated by MCL1 and FBW7. Nature. 2011;471(7336):110–4.
11. Zeng X, et al. Pharmacologic inhibition of the anaphase-promoting complex induces a spindle checkpoint-dependent mitotic arrest in the absence of spindle damage. Cancer Cell. 2010;18(4): 382–95.

Index

H. Inuzuka and W. Wei, *SCF and APC E3 Ubiquitin Ligases in Tumorigenesis*,   117
SpringerBriefs in Cancer Research, DOI 10.1007/978-3-319-05026-3,
© The Author(s) 2014